Study & Master

BIOLOGY

Grade 11
New Syllabus

N P J van Rensburg
J P van Wyk
J S Roux

SAYDBUC
We Help you Grow
South African Youth Development
and Business Chamber

The African Branch of Cambridge University Press supports
the South African Youth Development and Business Chamber
in youth development in South Africa

ROEDURICO

CAMBRIDGE UNIVERSITY PRESS
Cambridge, New York, Melbourne, Madrid, Cape Town, Singapore, São Paulo

Cambridge University Press
The Water Club, Beach Road, Granger Bay, Cape Town 8005, South Africa

www.cambridge.org
Information on this title: www.cambridge.org/9780947465346

© Cambridge University Press 2004

This book is in copyright. Subject to statutory exception and to the provisions of relevant collective licensing agreements, no reproduction of any part may take place without the written permission of Cambridge University Press.

First published 1996
Fourteenth impression 2007

Printed in South Africa by ABC Press

Cover design: The Graphic Shop
Illustrations: Arno Burger and Sunita Joyce

ISBN-13 978-0-947-46534-6 paperback
ISBN-10 0-947-46534-0 paperback

....

If you want to know more about this book or any other Cambridge University Press publication, phone us at (021) 412-7800, fax us at (021) 419-8418 or send an e-mail to capetown@cambridge.org

Preface

This book has been written to help reinforce and consolidate the basic biological knowledge required by the syllabus and to satisfy the needs of learners working towards the Senior Certificate Examination. It is designed to provide a source of material for classwork, homework and to serve as a revision programme. The overall aim of this book is to simplify the learner's task in acquiring the knowledge and understanding required by the Senior Secondary syllabus. For the teacher it provides a source of material that can be used for continuous evaluation and examining of learners.

The **subject-matter** in each chapter is presented in a direct, simple and factual way. To facilitate learning, the main sections are divided into **shorter sub-sections**. The diagrams, graphs and practical work incorporated into this book will not only enhance the learner's knowledge of the subject but also assist them in drawing meaningful conclusions. The **questions** include multiple choice, objective, structured, paragraph and for the Higher Grade a variety of data-response questions. They are designed to develop scientific skills, such as observation, interpretation and application of data. **Answers** to questions are given at the end of each chapter. It is recommended that learners should attempt to answer as many questions as possible on their own before referring to the answers at the end of the chapter. **Higher Grade** learners are recommended to work through **all the sections** of the book.

To the Learner

This is a learner-friendly book written specially for you to help you come to grips with the requirements of the syllabus. It is important that you use this book for independent self-study.

Biology is a key-subject in the school curriculum and if you use this book in the correct disciplined way, you will surely reap the benefits. You will also, simultaneously, develop a greater sense of responsibility, more self-discipline as well as greater self-motivation.

- **Read** the relevant sections in the chapter, that has been dealt with, attentively.
- Summarize, in **point-note form**, the subject-matter that was dealt with in class.
- Test yourself by **drawing the diagrams** and adding the labels.
- **Check** the accuracy of your diagrams and labels by comparing them with those in the book.
- **Answer** the questions at the end of each chapter. Under no circumstances should you read the answers before working through the questions yourself first.
- **Evaluate** your own work by comparing your answers with those in the book.
- Should **problems** arise, consult your teacher.

Best of luck! We hope this book will bring about positive change.

The Authors

Study & Master series

Grade 8	General Science
	Mathematics
Grade 9	General Science
	Mathematics
Grade 10	Accounting
	Biology
	Mathematics
	Physical Science
Grade 11	Biology
Grade 12	Biology
Grade 11 and 12 (*one book*)	
	Accounting
	Agricultural Science
	Mathematics *Higher Grade*
	Mathematics *Standard Grade*
	Physical Science

8 in one: Matric Question Papers
Accounting; Biology; Physical Science and Mathematics (SG & HG)
2003   2002 Senior Certificate Examination question papers and memorandums

Curriculum 2005

Grade 7	Journeying into Natural Sciences
Grade 8	Journeying into Natural Sciences
Grade 9	Journeying into Natural Sciences

Tips for Teens: Life Orientation
Survive your Life
Survive your Studies

RNCS

Grade 4	Stepping into Natural Sciences and Technology
Grade 5	Stepping into Natural Sciences and Technology
Grade 6	Stepping into Natural Sciences and Technology
Grade 4	Stepping into Mathematics
Grade 5	Stepping into Mathematics
Grade 6	Stepping into Mathematics

Tel: 021-852 3104 | Fax: 021-851 6874

Contents

1. Viruses, Bacteria, Mycophytes and Phycophytes — 1
2. Bryophytes and Pteridophytes — 24
3. Spermatophytes – Gymnosperms and Angiosperms — 50
4. Invertebrates – Phyla Protozoa and Coelenterata — 78
5. Invertebrates – Phyla Platyhelminthes and Annelida — 93
6. Invertebrates – Phylum Arthropoda — 109
7. Vertebrates – Phylum Chordata Subphylum Vertebrata — 124
8. Role of the Nucleus and Cell Division — 149
9. Genetic Mechanisms — 176
10. Reproduction in Human's — 193

1 Viruses, Bacteria, Mycophytes and Phycophytes

A. Viruses

Viruses are very small, acellular, non-living infectious organisms which have an impact on all living organisms on earth.

1. Viruses are **acellular** and can hardly be regarded as a form of life. They are often considered to be on the **border between living and non-living**. A virus consists of a core of either **DNA** or **RNA** (6%), surrounded by a **coat of protein** (94%) called a **capsid**. The DNA or RNA may be a single- or double-stranded molecule. Viruses that affect animals usually have DNA, and those that attack plants, RNA. An individual virus is called a **virion**. It has **no nucleus** or **cytoplasm**. Viruses vary in **shape**. Some are rod-shaped or spherical, but the most common is a 20-sided (icosahedron), almost round shape.

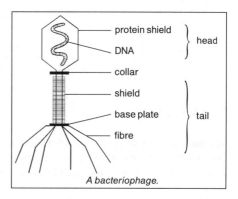

A bacteriophage.

2. The **biological importance** of viruses can be summarised as follows:
   (i) They all are **obligate parasites** and cause a variety of infectious diseases in plants and animals in which they live.
   (ii) During infection they **reproduce within host cells** by transforming the host nucleic acid into their own type and kill host cells.
   (iii) They are major pathogens of man (polio, smallpox, yellow fever, AIDS, influenza, measles), farm and domestic animals (foot-and-mouth disease, rabies) and crops (leaf mosaic disease). A pathogen is a parasitic organism which causes or produces a disease.
   (iv) It is difficult **to control** viruses since they are found inside host cells. Antibiotics cannot be used and vaccines are not always effective because one virus may exist in a variety of forms. Methods of control therefore depend primarily on **prevention**.
   (v) They can be **passed from** one person to another by direct contact, sneezing, coughing or disease vector.

B. Bacteria

Bacteria are the smallest **unicellular** and most plentiful organisms known.

1. Bacteria differ from all other kinds of living organisms in being **prokaryotic**. A prokaryote is an unicellular organism that has **no nucleus** or other membrane-bound **organelle**. Each cell is

surrounded by a **cell wall** and a **plasma membrane** just inside the cell wall. Outside the cell wall is a **slime capsule**. Many bacteria have **flagella** and can swim. Bacteria are found in variety of shapes and are classified as **bacilli**, **cocci**, **vibrios** and **spirilli**.

2. Most bacteria reproduce asexually by way of **binary fission**. The parent cell divides into two genetically identical daughter cells following DNA replication. In favourable conditions a bacterium cell can divide once every 20 minutes.

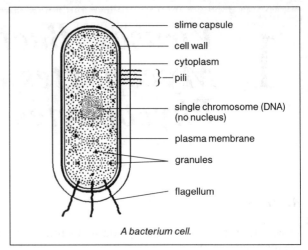

A bacterium cell.

3. The **ecological role** of bacteria can be summarised as follows:
   (i) They are found throughout the world and many are **decomposers**.
   (ii) They are important in the **nitrogen cycle** for nitrification, nitrogen fixation and denitrification.
   (iii) Some are **pathogens** and are spread by air, food, water, vectors and sexual contact.
   (iv) They play an important role as **symbionts** in animal nutrition since they live symbiotically in the alimentary canal of mammals.

C. Mycophytes

Mycophytes are **fungi** which include bread moulds, mushrooms and yeast. Fungi are eukaryotic organisms and have nuclei and membrane-bound organelles. They do not prossess chlorophyll, the green pigment, that is characteristic of most plants.

1. The bread mould, e.g. *Rhizopus* is **terrestrial** and grows on **moist** organic substrates in **warm**, **dark** places. *Rhizopus* commonly occurs in the soil and on food such as bread, cake and jam.

2. The fungal hyphae together form a **mycelium** with cell walls of **fungus-chitin**. The hyphae are non-septate as there are no true transverse walls and are **coenocytic**. **Stolons** form a branched covering over the surface of the substrate, **rhizoids** grow into the substrate and **sporangiophores** into the air.

3. *Rhizopus* lacks chlorophyll and the nutrition is **heterotrophic**. They live on non-living organic matter and are therefore **saprophytic**.

4. **Reproduction** is often associated with survival over unfavourable periods. Reproduction in *Rhizopus* can be **asexual** by haploid **spores** and **sexual** by fusion of nuclei which results in diploid zygospores.

5. Fungi play an important ecological role in nutrient cycling during decay. As **decomposers** they form an important link in food chains.

6. The **economic importance** of mycophytes can be summarised as follows:
   (i) Some fungi are pathogens and cause **diseases** like smuts, rusts, wilt and potato blight in agricultural crops; ringworm and athlete's foot are caused by parasitic fungi.
   (ii) During **fermentation** of yeast, alcohol (beer and wine production) and carbon dioxide (baking) are produced.

(iii) Some fungi are used to produce antibiotics, such as penicillin and cephalosporin.
(iv) Some fungi serve as **food** for human consumption. Certain mushrooms, e.g. *Agaricus bisporus* and *Agaricus brunnescens* are edible. Some mushrooms have hallucinogenic or deadly effects, e.g. *Amanita muscaria* (the fly agaric mushroom) and *Amanita phalloides* produce deadly toxins and must be avoided at all times.
(v) Some fungi decompose **stored foods**, like maize and wheat. The fungi release poisons called **mycotoxins** and the food becomes mouldy.

D. Phycophytes

Phycophytes are **algae**, of which *Spirogyra* is a common example. They range from small unicellular plants, like *Chlorella* to large multicellular types, as seaweeds and kelps.

1. Most phycophytes are adapted to an **aquatic** way of life. *Spirogyra* mainly occurs in shallow, stagnant, fresh water ponds.

2. *Spirogyra* is a green, free-floating, filamentous, fresh water alga. The plant body is a thallus because there are no specialised roots, stems or leaves.

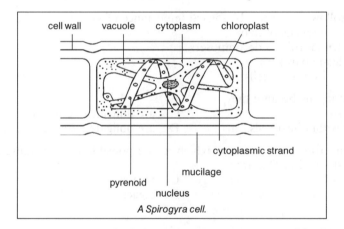

A *Spirogyra* cell.

The cells of *Spirogyra* are identical in structure and each cell contains a **nucleus** and **cytoplasm** enclosed by a cellulose cell wall. On the helical, ribbon-like **chloroplast** are **pyrenoids**.

3. Algae contain chlorophyll, can photosynthesise and are therefore **autotrophic**. Nutrients from the surrounding water medium diffuse directly through the thin cell wall.

4. **Reproduction** in *Spirogyra* may be **asexual** by **fragmentation** during favourable growth conditions or **sexual** by fusion of gametes when conditions become unfavourable. Sexual reproduction may be scalariform or lateral **conjugation**. The fusion of gametes results in a thick-walled **zygospore**. The haploid condition being restored when the zygospore nucleus undergoes **reduction division** (meiosis).

5. The **ecological role** of algae can be summarised as follows:
(i) Phytoplankton are **producers** and play an important role in the food chains of marine and fresh water animals.
(ii) They assist in maintaining the **oxygen and carbon dioxide balance** in an aquatic habitat.

STANDARD GRADE
QUESTIONS

Section A

A. *Various possibilities are suggested as answers to the following questions. Indicate the correct answer.*

1. Which of the following does *not* describe a virus? A virus
   A is an obligate parasite.
   B reproduces inside living cells.
   C consists of either DNA or RNA.
   D has a nucleus surrounded by a protein coat.

2. Viruses are mainly composed of
   A protein and nucleic acid.   C DNA and RNA.
   B cellulose and protein.   D cytoplasm and nuclei.

3. All viruses are
   A unicellular and pathogens   C acellular and non-living.
   B prokaryotes.   D cellular in structure.

4. Which of the following is *not* a biological importance of viruses?
   A Are major pathogens of man
   B Play an essential role as decomposers
   C All are obligate parasites
   D Reproduce within host cells

5. Viruses are pathogens because they
   A cause diseases.   C are difficult to control.
   B reproduce within host cells. D are obligate parasites.

6. The acellular microscopic organisms which are responsible for spreading diseases, such as influenza, measles and AIDS, are
   A fungi.   C viruses.
   B bacteria.   D algae.

7. Viruses
   A occur in water only.   C cause diseases.
   B are living unicellular organisms. D cause the decay of food.

8. Which of the following is *not* applicable to viruses? They
   A cause diseases.
   B are the simplest known organisms.
   C have the ability to reproduce themselves.
   D are living cells which are obligate parasites.

9. An outstanding characteristic of bacteria is that they
   A usually contain chlorophyll.
   B possess no nuclear material.
   C need light for growth.
   D possess a cellulose cell wall.

10. Which of the following is *not* a characteristic of bacteria?
    A Some are autotrophic and others saprophytic
    B The nucleus is surrounded by a nuclear membrane
    C They possess a cell wall
    D They are unicellular prokaryotes

11. Bacteria are pathogens because they
    A   feed on dead organisms.
    B   are prokarotic organisms.
    C   live symbiotic in the intestine of mammals.
    D   cause diseases.

12. A bacterium cell
    A   is a prokaryote.
    B   has a nuclear membrane around its genetic material.
    C   is a eukaryote.
    D   contains mitochondria, vacuoles and plastids in the cytoplasm.

13. Bacteria reproduce
    A   sexually only.   C   inside living host cells only.
    B   by binary fission.   D   by fragmentation.

14. Binary fission is a method of reproduction that generally occurs in
    A   bacteria.   C   fungi.
    B   viruses and bacteria.   D   fungi and algae.

15. Which of the following statements with regard to bacteria and humans is true?
    A   All bacteria are pathogens
    B   All bacteria are useful
    C   Useful bacteria are sometimes harmful
    D   Bacteria reproduce only within a body cell

16. The correct sequence of processes of the nitrogen cycle in nature.
    A   Biosynthesis   Decomposition   Nitrification   Ammonification
    B   Ammonification Biosynthesis   Nitrification   Decomposition
    C   Biosynthesis   Decomposition   Ammonification   Nitrification
    D   Nitrification   Ammonification   Decomposition   Biosynthesis

17. Ringworm and athlete's foot are caused by
    A   bacteria.   C   viruses.
    B   fungi.   D   algae.

18. A saprophyte is
    A   capable of autotrophic nutrition.
    B   a parasite.
    C   incapable of synthesising its own organic food.
    D   always dependent on an animal for its food.

19. In the life cycle of *Rhizopus* the process of meiosis occurs
    A   in the sporangium at the tip of the promycelium.
    B   just before the zygospore ruptures.
    C   during the formation of the gametangia.
    D   during the production of gametes.

20. A typical habitat of *Rhizopus* is
    A   slow, flowing freshwater.   C   the skin of most animals.
    B   damp, shady soil.   D   non-living, organic substances.

21. The promycelium of *Rhizopus* develops from the
    A   zygospore.   C   sporangiophore.
    B   sporangium.   D   sporogonium.

Questions 22 and 23 are based on the diagram below.

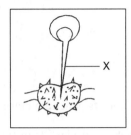

22. To which of the following groups of plants does the structure in the diagram above belong?
    A   Bryophytes
    B   Phycophytes
    C   Pteridophytes
    D   Mycophytes

23. X refers to the
    A   promycelium.
    B   sporangium.
    C   plus strain.
    D   zygospore.

24. The method of rapid reproduction in *Spirogyra* when conditions are favourable for growth.
    A   Sexual
    B   Fragmentation
    C   Conjugation
    D   Spore formation

25. A thallus plant without chlorophyll occurs in
    A   phycophytes.
    B   pteridophytes.
    C   mycophytes.
    D   bryophytes.

26. Which of the following mode of nutrition is indicated by the presence of chloroplasts in leaves?
    A   Saprophytism
    B   Autotrophic
    C   Heterotrophic
    D   Parasitism

27. Which of the following organisms possesses pyrenoids?
    A   Bacteria
    B   Plants with chloroplasts
    C   Algae
    D   Fungi

28. The pyrenoids in *Spirogyra* are concerned with
    A   conjugation.
    B   starch storage.
    C   cell division.
    D   photosynthesis.

29. A sporangium produces
    A   spores.
    B   zygospores.
    C   sperm cells.
    D   eggs cells.

Questions 30 and 31 refer to the following diagram.

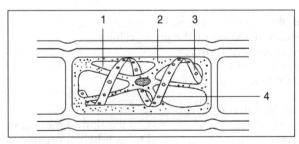

30. Which numbered part will turn blue-black in colour in contact with diluted iodine solution after chlorophyll has been extracted?
    A   1
    B   2
    C   3
    D   4

31. Which numbered part indicates that it is an autotrophic organism?
  A 1
  B 2
  C 3
  D 4

32. The fusion of a sperm cell and an ovum is called
  A asexual reproduction.
  B vegetative reproduction.
  C fertilisation.
  D conjugation.

33. Which of the following could be concerned with asexual reproduction?
  A Spores
  B Gametes
  C Zygotes
  D Ovaries

34. The cell walls of most fungi are mainly composed of
  A cellulose.
  B chitin.
  C pectin.
  D lignin.

35. Which of the following is the *most* important role of both bacteria and fungi in a community of living organisms?
  A Spreading of disease
  B Fixation of nitrogen
  C Alcohol fermentation
  D Decomposition of organic materials

36. A filamentous alga is classified as a plant because it
  A is differentiated into roots, stems and leaves.
  B is immobile.
  C reproduces sexually and asexually.
  D possesses chloroplasts.

37. When growth conditions become unfavourable, *Spirogyra* reproduces by
  A the formation of gametes.
  B fragmentation.
  C the formation of haploid spores.
  D binary fission.

38. AIDS is a deadly disease which can be spread
  A by bacteria.
  B during sexual intercourse.
  C in the air.
  D by vectors.

39. The micro-organisms which are responsible for most diseases and infections in man are:
  A bacteria and viruses.
  B viruses.
  C bacteria and fungi.
  D fungi.

40. Fungi obtain their energy from
  A sunlight.
  B alcohol.
  C organic matter.
  D yeast.

B. Write down the correct term for each of the following statements.

1. An organism that consists of a nucleic acid surrounded by a shield of protein
2. A substance which is found in the central part of a virus
3. Disease-causing bodies smaller than bacteria that can only reproduce inside living cells
4. An unicellular organism that is neither plant nor animal
5. Organisms like bacteria in which cells are characterised by having no true nuclei
6. A disease-causing parasite, e.g. bacteria
7. The type of asexual reproduction whereby unicellular organisms like bacteria, divide approximately into two equal parts
8. The place where an organism normally lives and interacts with the abiotic and biotic environments
9. The habitat which describes living on land

10. The habitat which describes living in water
11. A nutritional relationship in which two different organisms live together
12. The mass of hyphae that constitutes the vegetative part of a fungus
13. A plant body which is not differentiated in roots, stems and leaves
14. A mode of nutrition where only inorganic substances are taken in to synthesise organic food
15. The body of the fungus
16. The process by which ammonia and nitrites are converted into nitrates by bacteria
17. The small structures on the chloroplast in which organic food is stored in many algae
18. The group of plant to which plants, having a single large chloroplast, belong
19. The vegetative mode of reproduction in *Spirogyra*
20. The dome-shaped cross wall in the sporangium of *Rhizopus*
21. The collective name given to the microscopic small free-floating plant forms of plankton found in the upper layers of water
22. A mode of nutrition where an organism is incapable of synthesising organic compounds from raw materials, therefore depending on other organisms for food
23. A close association between a fungus and the tissue of a root
24. A cell resulting from sexual fusion of two gametes
25. The sexual reproduction resulting from the fusion of gametes of similar size and appearance
26. A cell, tissue or organ in mycophytes in which gametes are produced
27. The name of the mutualistic bacteria found in the nodules of legumes

C. *Write down the letter of the description in column B which best suits the term in column A.*

Column A		Column B
1. Symbiosis	A	Method of nutrition in algae
2. Heterotrophic	B	Thallus plants are morphologically similar but physiologically different
3. Zygospore		
4. Saprophytic	C	Spermatophytes produce two types of spores
5. Homothallic	D	Referring to species in which male and female gametangia are produced in the same filament
6. Autotrophic		
7. Heterosporous	E	Possess no chlorophyll to produce organic food substances
8. Heterothallic	F	Method of nutrition in fungi
	G	Bacteria on roots of legumes
	H	Zygote with resistant wall

D. *Study the following diagrams and answer the questions*

1. (a) Name the group of plants to which this plant belongs.
   (b) What type of reproduction is shown here?
   (c) Identify the parts numbered A and B.

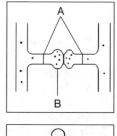

2. (a) Identify the parts numbered A and B.
   (b) Name the plant represented by this diagram.

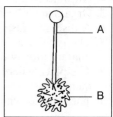

3. (*a*) Name the group of plants to which this plant belongs.
   (*b*) What type of reproduction is shown here?
   (*c*) Identify the parts numbered A and B.

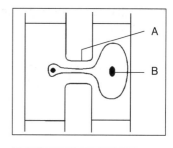

4. (*a*) Identify the parts numbered A and B.
   (*b*) Name the plant represented by this diagram.

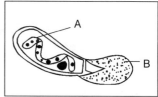

Section B

1. Briefly describe the structure of viruses.

2. (*a*) Briefly describe why viruses are always harmful.
   (*b*) In what ways can viruses be dispersed?

3. (*a*) Why are bacteria said to be prokaryotic organisms?
   (*b*) State the **four** main shapes in which bacteria can be distinguished.
   (*c*) Draw a labelled diagram to illustrate the structure of a bacterium cell.

4. (*a*) Under which growth conditions reproduce bacteria asexually?
   (*b*) Briefly describe the reproduction of a bacterium cell under the conditions mentioned in (*a*).

5. Bacteria as pathogens can enter the human body in several ways. Briefly describe these ways and state in each case the diseases carried this way and how it can be controlled.

6. Briefly describe the ecological role of bacteria.

7. Complete the following diagram of the nitrogen cycle in nature by writing down the missing word(s) numbered 1 to 12.

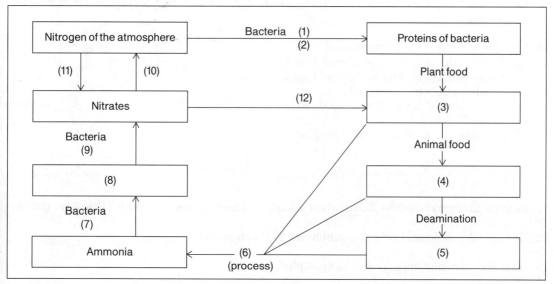

8. The diagram below represents the life cycle of the fungus you have studied. Answer the following questions.

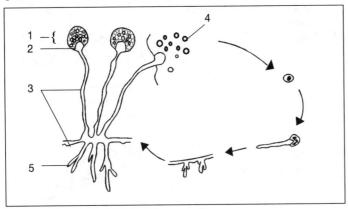

   (a) Name the group of plants to which this plant belongs.
   (b) Identify the part numbered 1.
   (c) Write down the number which represents the only cross wall developed in this plant. What is this cross wall called?
   (d) Identify the part numbered 4. How is it released and dispersed?
   (e) What type of reproduction is shown here?
   (f) Identify structure numbered 5. Where is it growing? How does it obtain food? Is this an example of extracellular or intracellular digestion?
   (g) Is this plant a parasite or saprophyte?
   (h) What colouring matter or pigment is this plant lacking, which is found in a normal flowering plant, e.g. a sweet pea?

9. Answer the following questions on the fungus you have studied.
   (a) Give the name of the fungus you have studied.
   (b) Why can this fungus be described as a thallus plant?
   (c) Give a reason why this plant can be classified as a saprophyte.
   (d) Give a reason why the mycelium of this plant is referred to as a coenocyte.
   (e) Mention **three** environmental conditions in which this fungus flourishes.
   (f) Mention the stage in the life cycle of this fungus that is diploid.
   (g) Under which growth condition and by means of which process does this fungus reproduce asexually?

10. This diagram represents a fungus.
    (a) Identify the structures numbered 1 to 3.
    (b) Identify structure numbered 4 and state its function.
    (c) Identify structure numbered 5.
    (d) What form of heterotrophic mode of nitrition is found in the fungus?
    (e) Describe the nutrition of this fungus.

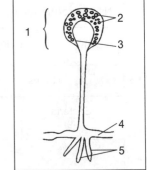

11. Describe the reproduction of *Rhizopus* when conditions become unfavourable for vegetative growth.

12. Describe the ecological role of mycophytes as decomposers.

13. State **six** economic importances of mycophyta.

14. The diagram represents *Spirogyra*.
    (a) Name the group of plant to which this plant belongs.
    (b) Identify the parts numbered 1 to 8.
    (c) State **one** function of part numbered (i) 4, and (ii) 5.
    (d) Does part numbered 6 contain a haploid or diploid number of chromosomes?
    (e) Mention **three** enviromental conditions in which *Spirogyra* flourishes.

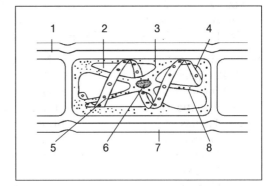

15. Briefly describe the
    (a) habitat, and
    (b) nutrition of algae.

16. Describe asexual reproduction in *Spirogyra*.

17. The diagram below shows stages during reproduction in *Spirogyra*.

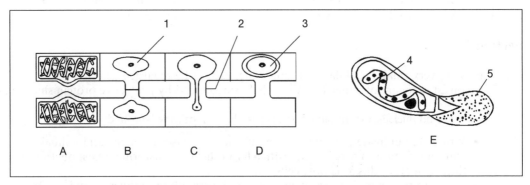

    (a) What type of reproduction is illustrated by stages numbered A to D?
    (b) Mention **two** environmental conditions most likely to serve as a stimulus for the type of reproduction mentioned in (a)
    (c) Identify the parts numbered 1 to 5.
    (d) Briefly describe the events which occur during stage C.
    (e) What type of reproduction is illustrated by stage E?
    (f) Mention **three** environmental conditions most likely to serve as a stimulus for the type of reproduction mentioned in (e).

18. Briefly describe the ecological role of algae
    (a) as producers in ecosystems, and
    (b) in maintenance of oxygen – carbon dioxide balance.

STANDARD GRADE
ANSWERS

Section A

A. 1. D   2. A   3. C   4. B   5. A   6. C   7. C
    8. D   9. D   10. B 11. D 12. A 13. B 14. A
    15. C 16. C 17. B 18. C 19. B 20. D 21. A

22. D	23. A	24. B	25. C	26. B	27. C	28. B
29. A	30. D	31. C	32. C	33. A	34. B	35. D
36. D	37. A	38. B	39. A	40. C		

B. 1. Virus 2. Nucleic acid 3. Viruses
4. Bacterium 5. Prokaryotes 6. Pathogen
7. Binary fission 8. Habitat 9. Terrestrial
10. Aquatic 11. Symbiosis 12. Mycelium
13. Thallus 14. Autotrophic 15. Mycelium
16. Nitrification 17. Pyrenoids 18. Phycophytes
19. Fragmentation 20. Columella 21. Phytoplankton
22. Heterotrophic 23. Mycorrhiza 24. Zygote
25. Conjugation 26. Gametangium 27. *Rhizobium*

C. 1. G 2. E 3. H 4. F 5. D 6. A 7. C 8. B

D. 1. (*a*) Mycophytes (*b*) Sexual (*c*) A – cross walls B – gametangium
2. (*a*) A – promycelium B – zygospore (*b*) *Rhizopus*
3. (*a*) Phycophytes (*b*) Sexual (*c*) A – conjugation tube B – nucleus
4. (*a*) A – chloroplast B – zygospore (*b*) *Spirogyra*

Section B

1. Viruses are very small, acellular and non-living; cannot be regarded as cells
   They consist of a core of either DNA or RNA; surrounded by protective protein shield called capsid
   Viruses have no nucleus, cytoplasm or organelles; their shape may be round, rod-shape or spherical

2. (*a*) • Viruses are obligate parasites; cause infectious diseases in plants and animals
   • During infection they reproduce **within host cells**; by transforming host nucleic acid into their own type; this kills host cells
   • They are major **pathogens**; of man like polio, smallpox, influenza and AIDS; and of plants like leaf mosaic disease
   • Viruses are **difficult to control** since they are found inside host cells; antibiotics do not help and vaccines are not always effective
   (*b*) Direct contact; sneezing; coughing; disease vectors

3. (*a*) Bacteria are unicellular organisms; with no nucleus or other membrane-bound organelles
   (*b*) Bacillus (rod-shaped); coccus (round); vibrio (comma-shaped); spirillum (spiral-shaped)
   (*c*)

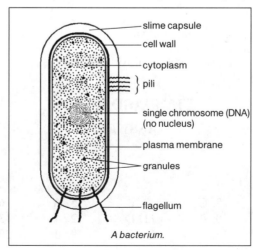

A bacterium.

4. (*a*) Under favourable conditions; i.e. sufficient food and water; optimum temperature and darkness
   (*b*) Bacteria reproduce asexually by binary fission; chromosome divides into two; each part containing a copy of the DNA; daughter DNA attaches at a position near site of parent DNA molecule; plasma membrane and cell wall start to grow between; two attachment sites and move DNA molecules apart; plasma membrane and cell wall grow through midsection; cytoplasma divides into two; each cell grows to its maximum size; process is a rapid one and can take place every 20-30 minutes.

5. Through the **air**
   - By coughing, sneezing or spitting
   - tuberculosis, pneumonia and diptheria
   - notify authorities and obtain medical treatment.

   In **food** and **water**
   - If food has been in contact with faeces from people suffering from a disease
   - flies spread bacteria from faeces to food
   - faeces may contaminate water supplies
   - botulism, gastro-enteritis and cholera
   - keep water supply clean and boil drinking water.

   Through the **skin**
   - Bacteria can enter body through cuts and grazes
   - tetanus, boils, abscesses and impetigo
   - regular cleaning of skin and disinfecting of wounds.

   By **disease vectors**
   - Rat flea carries bacteria
   - bubonic plague
   - exterminate rats in populous regions.

   By **sexual contact**
   - Serious diseases can be passed from one person to another during sexual intercourse
   - venereal diseases like syphilis and gonorrhoea
   - sex education.

6. Role as **decomposers**
   - They form an important link in food chains
   - decompose bodies of dead plants and animals
   - convert organic compounds into inorganic substances
   - which replenish the nutrients needed by plants and animals.

   Role in **nitogen cycling**
   - All living organisms need nitrogen
   - which is constituent of proteins and nucleic acids
   - organisms cannnot capture nitrogen directly from the air
   - nitrogen-fixing bacteria convert nitrogen into ammonia
   - which disssolves to form soluble nitrogen salts
   - absorbed by plants which provide animals with amino acids.

   Role as **pathogens**
   (refer to previous question)

   Role as **symbionts**
   - Some bacteria live symbiotically in gut of animals
   - Where they digest cellulose and make nitrogen available
   - *Rhizobium* has a symbiotic relationship with legumes
   - in the nodules they fix nitrogen gas for plant consumption.

7. 1 Free living nitrogen-fixing bacteria/*Azotobacter*/*Clostridium*
   2 Symbiotic/*Rhizobium*   3 Plant protein   4 Animal protein
   5 Urea   6 Ammonification   7 *Nitrosomonas*/*Nitrosococcus*
   8 Nitrites   9 *Nitrobacter*   10 Denitrification
   11 Electrical fixation   12 Absorbed by plants

8. (*a*) Mycophytes   (*b*) Sporangium   (*c*) No. 2 – columella
   (*d*) Spore • Sporangium wall dries out
         • columella pushes deeper into sporangium
         • or wall absorbs water to become soft
         • columella exerts pressure on wall
         • wall bursts and spores are released
         • spores are light and are dispersed by wind.
   (*e*) Asexual
   (*f*) Rhizoid; • In bread or moist organic substances;
                  • secrete diastase that converts starch into glucose;
                  • nitrates and phosphates are absorbed directly;
                  Extracellular
   (*g*) Saprophyte   (*h*) Chlorophyll

9. (*a*) *Rhizopus*
   (*b*) Plant body is not differentiated into roots, stems and leaves
   (*c*) Food is absorbed from non-living organic matter, such as bread/jam
   (*d*) Hyphae are non-septate; protoplasm is continuous
   (*e*) Moist, warm, dark places
   (*f*) Zygote
   (*g*) Favourable; spore formation

10. (*a*) 1 – Sporangium   2 – Spores   3 – Columella
    (*b*) Stolon • Forms rhizoids downwards into substrate for absorption of food and water;
          • and sporangiophores upwards to produce sporangia.
    (*c*) Rhizoids   (*d*) Saprophytic
    (*e*) Posesses no chlorophyll and can not photosynthesise; lives on non-living, moist organic matter, e.g. bread; rhizoids penetrate bread and secrete enzyme diastase; which converts starch into glucose; glucose is absorbed together with nitrates, phosphates, water and other mineral salts; to form required proteins and other organic substances.

11. **Sexual** reproduction occurs when it becomes dry and temperature drops
    Two hyphae of two **different strains** (+ and – strain) grow towards each other
    The tips become **swollen** due to dense cytoplasm with many nuclei
    Progametangia are formed
    The progametangia are separated from **suspensors** by crosswalls
    Gametangia are formed
    Wall between gametangia dissolves and many + and – nuclei **pair off** and **fuse**
    To form a **zygote** with many diploid nuclei
    The zygote develops a hard thick wall and becomes a **zygospore**
    Zygospore becomes **detached** from parent hyphae
    In **dormant state** it can withstand unfavourable growth conditions.

    > After a period of rest when **conditions become favourable** the zygospore develops
    > All **but one** of diploid nuclei break down
    > Remaining nucleus divides by **meiosis** to give rise to 4 haploid cells
    > Three of which also break down and remaining one may be + or –
    > Outer wall of zygospore bursts and **promycelium** emerges
    > Haploid nuclei divide **mitotically** in the sporangium
    > Nuclei together with cytoplasm form **spores**
    > Which are dispersed by wind to give rise to new **haploid mycelium**

12. Form an important link in food chains
    Cause decay of organic remains of plants and animals
    Inorganic compounds derived from decay are released and recycled in nature
    In this way nutrients are returned to soil and atmosphere
    To be re-used by plants and animals
    Decay process involves a sequence of fungal species
    Breakdown of organic substrate starts with those fungi which decompose sugars
    Followed by those causing breakdown of starch
    Then those breaking down cellulose and finally lignin
    Earlier stages take place rapidly and later stages slowly

13. Some are serious pathogens of plants and animals
    Some are used as food for human consumption
    Some are used in production of alcohol
    Some are used in production of food
    Some are used to produce antibiotics
    Some decompose stored foods

14. (a) Phycophytes
    (b) 1 Cell wall   2 Vacuole   3 Cytoplasm
        4 Chloroplast   5 Pyrenoid   6 Nucleus
        7 Mucilage   8 Cytoplasmic strand
    (c) (i) Contains chlorophyll and photosynthesises
        (ii) Stores produced organic food mainly starch
    (d) Haploid
    (e) • Optimum temperature
        • Shallow, stagnant fresh water
        • Sufficient sunlight

15. (a) Shallow stagnant fresh water
    (b) Ribbon-like chloroplast contains chlorophyll and photosynthesise
        They are autotrophic
        CO_2 diffuses from surrounding water into cell
        Using light energy and water to convert it into glucose and starch
        Stored in pyrenoids
        Absorbs mineral salts
        And proteins are synthesised

16. When filament breaks
    Due to water currents or swimming animals
    Each part of filament divides mitotically
    To form new independent filament
    This is called fragmentation
    Under favourable growth conditions, i.e.
    Sufficient water and food and suitable temperature
    Nucleus divides mitotically into two equal parts
    And move to opposite sides of parent cell
    Ringlike invagination divides wall of parent cell into two
    Two cells eventually grow to adult size and separate.

17. (a) Sexual   (b) Drop in temperature; It becomes dry;
    (c) 1 – Female gamete   2 – Conjugation tube   3 – Zygote
        4 – Chloroplast   5 – Zygospore
    (d) • Male gamete moves through conjugation tube
        • And conjugates with stationary gamete
        • Stationary gamete is the female gamete
        • Fusion results in a diploid zygote

(e) Asexual or vegatative
(f) • Optimum temperature
    • Shallow, stagnant fresh water
    • Sufficient sunlight.

18. (a) Play important role in food chain of marine and fresh water ecosystems
    They contain chlorophyll
    Synthesise their own organic food by photosynthesis
    They provide food for zooplankton
    And indirectly for most other marine organisms
    (b) During photosynthesis carbon dioxide is absorbed
    To produce organic food like glucose and starch
    Oxygen is released to the atmosphere

HIGHER GRADE
QUESTIONS

Section A

A. Various possibilities are suggested as answers to the following questions. Indicate the correct answer.

1. A capsid in a virus is a
   A   method of reproduction.
   B   nucleic acid.
   C   protective protein coat.
   D   virion.

2. A biologist discovered a new living cell with a distinct cell wall but no definite nucleus. The cell is likely to be that of a/an
   A   animal.
   B   plant.
   C   virus.
   D   bacterium.

3. A pathogenic organism is one that
   A   causes decay.
   B   causes disease.
   C   combats disease germs.
   D   acts as antibiotics.

4. The two main structural features of viruses are a
   A   nuclei acid core and a protein coat.
   B   DNA-containing nucleus and a lipid envelope.
   C   nucleic acid core and a plasma membrane.
   D   DNA core and a protein coat.

5. Viruses infect
   A   bacteria.
   B   animals.
   C   plants.
   D   all of the above-mentioned.

6. Which of the following describes bacteria best?
   A   Unicellular plants, parasitic in plants and animals
   B   Unicellular, non-nucleated organism which multiply rapidly
   C   Microscopic plants with saprophitic mode of nutrition
   D   Unicellular, nucleated organisms parasitic in plants and animals

7. The nitrogen cycle in nature is vital because animals and most green plants cannot
   A   use gaseous nitrogen.
   B   use nitrates.
   C   use organic nitrogen compounds.
   D   synthesise their own proteins.

8. This drawing is of a bacterium.
   A feature here which is characteristic of **all** bacteria is that it

   A   has a flagellum
   B   has chromatin clumped in the centre.
   C   is enclosed by a cell wall.
   D   has a slime capsule on its outer surface.

9. Most phycophytes possess pyrenoids in which
   A   starch is stored.   C   glucose is oxidised.
   B   photosynthesis takes place.   D   protein synthesis takes place.

10. The hyphae of *Rhizopus* is coenocytic and is therefore
    A   enucleate.   C   binucleate.
    B   uninucleate.   D   multinucleate.

11. Reduction division in *Spirogyra* takes place
    A   during the germination of the zygospore. C   during the formation of the gametes.
    B   in the conjugation tube.   D   just before conjugation takes place.

12. A hypha is to a mycelium as a cell is to
    A   a cell wall.   C   a tissue.
    B   a nucleus.   D   a spore.

13. Each thread composing the structure of *Rhizopus*, is known as a
    A   sporangiophore.   C   mycelium.
    B   hypha.   D   stolon.

14. Which statement about binary fission in bacteria is incorrect?
    A   It is a sexual process
    B   It takes place when conditions are favourable for growth
    C   It is preceded by reduction division
    D   Colonies of bacteria are formed

15. In an ecosystem bacteria and fungi are
    A   predators.   C   decomposers.
    B   producers.   D   omnivores.

16. The vegetative part of a fungus is called a
    A   mycelium.   C   sporangium.
    B   sporangiophore.   D   spore.

17. In the structure of *Rhizopus* there is no clear division into cells as in *Spirogyra*. The plant body
    in *Rhizopus*, therefore, is described as
    A   coenocytic.   C   eucaryotic.
    B   multicellular.   D   heterothallic.

18. In the following drawing **X** refers to a

    A   endospore.
    B   sporangiophore.
    C   gamete.
    D   zygote.

19. Scalariform conjugation can be observed under the microscope by studying specimens of
    A   bacterium species.   C   *Spirogyra*.
    B   *Amoeba*.   D   *Rhizopus*.

17

20. The thick-walled dormant zygote which is a product of sexual reproduction in bread mould, is called a
    A gametangium.
    B zygote.
    C zygospore.
    D promycelium.

21. Which of the following is **not** a feature of fungi?
    A They are all saprophytic
    B They are all autotrophic
    C The vegetative part of the plant forms a mycelium
    D The fungal body is a thallus

22. Which of the following in *Spirogyra* is diploid?
    A Gametangium
    B Nucleus of parent filament
    C Nucleus of *Spirogyra* cell
    D Zygospore

23. Algae and fungi are similar in that
    A both are decomposers in ecosystems.
    B the plant body is a thallus.
    C both are autotrophic organisms.
    D sexual reproduction does not occur.

Questions 24 to 26 refer to the following diagram.

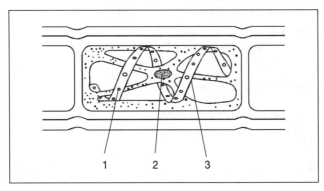

24. The structure numbered 1
    A contains chlorophyll for photosynthesis.
    B stores starch.
    C has a diploid number of chromosomes.
    D is the nucleus.

25. Which numbered part(s) make(s) it possible to synthesise organic substances from inorganic substances?
    A 1 only
    B 1 and 2
    C 2 only
    D 3 only

26. Which numbered part(s) is (are) responsible for controlling the metabolism of the cell?
    A 1 only
    B 1 and 2
    C 2 only
    D 3 only

27. Plants that are commonly called seaweeds, are likely to be
    A algae.
    B angiosperms.
    C lichens.
    D mosses.

28. Spores and gametes are both
    A formed during unfavourable conditions.
    B formed mitotically.
    C haploid.
    D diploid.

B. Write down the correct term for each of the following statements.

1. A protein which can be formed in animal tissue in response to a virus and inhibits the virus from multiplying
2. A separate and individual genetic unit in the form of a ring of DNA found in a bacterium
3. An animal, usually an insect, that carries and transmits pathogenic micro-organisms from one host to another host
4. A unicellular organism that can exist in coccus, spirillum or bacillus form
5. The name of the process for the conversion of ammonia and nitrites into nitrates by bacteria
6. An association of two different kinds of living organisms involving benefit to both
7. The structure which develops when a zygospore of *Rhizopus* germinates
8. The structure where starch is stored in a filamentous alga
9. The vegetative part of a fungus
10. The type of nutrition in plants having no chlorophyll
11. The term that indicates that *Rhizopus* has two types of mycelia which are morphologically identical but differ physiologically
12. A mass of cytoplasm, enclosed in a single continuous plasma membrane, with many nuclei
13. The name of the bacteria that reduce nitrogen gas from soil air into ammonia in root nodules of legumes
14. Cells which fuse to form a zygote
15. The protective coat of protein which encloses the nucleic acid in a virus
16. A stalk bearing a sporangium
17. Process of sexual reproduction involving the fusion of isogametes
18. Decomposers living heterotrophically on the remains of plants and animals except bacteria

Section B

1. (a) Name **three** kinds of nitogen-fixing bacteria in the soil and state where each is found.
   (b) What is the process called whereby proteins in the soil are changed by decomposing bacteria into ammonia?
   (c) By what kind of bacteria can ammonia be oxidised to nitrites? Give **two** examples of this kind of bacteria.
   (d) Which kind of bacteria can oxidise poisonous nitrites into nitrates? Give one example of this kind of bacteria.
   (e) Give the name of the process by which nitrates can be reduced to nitrogen gas.
   (f) Give the name of the process by which bacteria reproduce in favourable conditions.

2. (a) Name the fungus you have studied and briefly describe its general structure.
   (b) Answer the following questions with regard to nutrition of this fungus:
      1. Is this method of nutrition heterotrophic or autotrophic? Give a reason for your answer.
      2. Explain how organic nutrients are made available to the fungus.
      3. Name: (i) a disadvantage; and
               (ii) an advantage of this method of nutrition.

3. (a) Name: (i) a difference, and (ii) a similarity between the vegetative structure of phycophytes and mycophytes.
   (b) Give an example of a filamentous alga. How would you identify this plant under the microscope?
   (c) State a reason why one can assume that phycophytes contain chlorophyll although they are not all green.
   (d) (i) When does sexual reproduction usually occur in this alga?
       (ii) What are the advantages of this form of reproduction to this plant?

4. Explain how *Rhizopus* is well suited to terrestrial life (structure, nutrition and reproduction).

5. (*a*) Where does meiosis take place in *Spirogyra*?
   (*b*) Is the filament of *Spirogyra* haploid or diploid?
   (*c*) Draw a labelled diagram of the germinating zygospore of *Spirogyra*.

6. Study the following diagram of a phycophyte species and answer the questions.

   (*a*) Identify this example.
   (*b*) Identify the parts numbered 1 to 6.
   (*c*) Of which significance is the layer of mucus to this example?
   (*d*) Give reasons why sexual reproduction in algae can be regarded as a mechanism for survival.

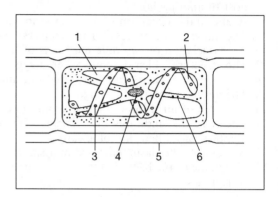

7. What is meant by: (*a*) thallus plants, and
                     (*b*) alternation of generations in plants?

8. Briefly describe the possible significance of the relationship (association) between the organism and tissue in each of the following cases:
   (*a*) *Rhizobium* in the root of a leguminous plant.
   (*b*) Non-pathogenic bacteria species in the rumen of a cow.
   (*c*) *Escherichia coli* in the colon of man.
   (*d*) Active *mycobacterium tuberculosis* in alveolar tissue of the lung.

9. (*a*) Name the example of a filamentous phycophyte you have studied.
   (*b*) Explain how the named example is structuraly well suited to survive in its particular habitat with regard to nutrition.

HIGHER GRADE
ANSWERS

Section A

A. 1. C   2. D   3. B   4. A   5. D   6. B   7. A
    8. C   9. A   10. D   11. A   12. C   13. B   14. A
   15. C   16. A   17. A   18. D   19. C   20. C   21. B
   22. D   23. B   24. B   25. D   26. C   27. A   28. C

B. 1. Interferon   2. Plasmid   3. Vector
    4. Bacterium   5. Nitrification   6. Mutualism
    7. Promycelium   8. Pyrenoid   9. Mycelium
   10. Heterotrophic   11. Heterothallic   12. Coenocyte
   13. *Rhizobium*   14. Gametes   15. Capsid
   16. Sporangiophore   17. Conjugation   18. Fungi

Section B

1. (a) *Azotobacter* – well aerated sweet topsoil
   Clostridium – anaerobic conditions in acid subsoil
   Rhizobium – in nodules on roots of legumes
   (b) Ammonification
   (c) Nitrite bacteria – *Nitrosomonas* and *Nitrosococcus*
   (d) Nitrate bacteria – *Nitrobacter*
   (e) Denitrification   (f) Binary fission

2. (a) *Rhizopus* sp.
   Plant thallus, consisting of hyphae, is the mycelium;
   which forms horizontal stolons;
   and sporangiophores vertically into the air;
   and rhizoids downwards into the substrate.
   Hyphae are non-septate with cell walls of fungus-chitin;
   thin layer of cytoplasm with numerous small haploid nuclei;
   protoplast contains large central vacuole with cell sap.
   On top of sporangiophore is the sporangium.
   Mycelium can be thought of as coenocyte.

   (b) 1 Heterotrophic – possesses no chlorophyll and can not photosynthesise
   2 Rhizoids grow into substrate; protoplasts of rhizoids secrete diastase; that converts starch into glucose; glucose is absorbed together with nitrates and phosphates in a watery solution; to form proteins; and other organic substances of fungal protoplasm.
   3 (i) Is dependent on non-living organic matter; when food source is exhausted, the plant dies.
   (ii) No complicated structures necessary for photosynthesis are required; food is absorbed in a prefabricated form.

3. (a) (i) Phycophytes possess chloroplasts with chlorophyll and mycophytes not
   (ii) Both are thallus plants
   (b) *Spirogyra*
   - plant body is a thallus
   - filament is non-septate with large number cylindrical cells
   - cellulose cell wall is coated with mucus layer
   - ribbon-like chloroplast on inner periphery
   - on broader parts of chloroplast are pyrenoids
   - centrally is nucleus supported by cytoplasmic strands.
   (c) Due to presence of the chloroplast
   (d) (i) At onset of unfavourable growth conditions;
       cold and dry.
       (ii) • Introduce variability as a result of genetic variation
       • zygospore is formed
       • with thick waterproof wall
       • serves as protection against unfavourable conditions
       • will only germinate at start of favourable growth conditions
       • in this way survival is ensured.

4. Hyphae grow in and over substrate
   Rhizoids penetrate substrate to absorb salts and water
   Cell wall consists of fungus-chitin giving strength and flexibility
   Rhizoids secrete diastase;
   which converts starch into glucose;
   absorbs salts and water from substrate
   Elongated sporangiophores ensure that sporangia are borne well above substrate;
   for effective spore dispersal by wind.
   Columella exerts pressure and wall bursts to disperse spores
   Spores are light and well suited to dispersal by wind.

5. (a) In the zygospore
   (b) Haploid
   (c)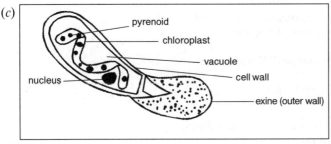

6. (a) *Spirogyra*
   (b) 1 – vacuole   2 – chloroplast   3 – pyrenoid
       4 – nucleus   5 – mucilage   6 – cytoplasmic strand
   (c) Mucilage layer causes filaments to stick to one another
       Close proximity permits easy conjugation
       Slippery nature reduces frictional damage between matted filaments
       Clumping of filaments has also nutritional advantage
       Bubbles cause filamentous mass to float near water surface;
       in favourable position for light absorption (photosynthesis).
   (d) Under unfavourable conditions for growth;
       sexual reproduction occurs.
       After zygote is formed a tough resistant waterproof wall develops.
       Zygospore can withstand extremes of temperature and is resistant to drying out
       It can survive for a long time in this dormant state.

7. (a) Plant body is not differentiated into roots, stems and leaves
   (b) That is a sexual life cycle;
       where a gamete producing generation;
       with a single set of chromosomes (haploid/n) in each nucleus,
       the gametophyte, alternating;
       with a spore producing generation;
       with a double set of chromosomes (diploid/2n) in each nucleus,
       the sporophyte.

8. These are examples of mutualistic associations
   (a) • Bacteria fix free nitrogen gas from soil air
       • and reduce it to ammonia
       • which becomes available to host plant
       • in exchange the bacteria are given a sheltered habitat
       • and obtain their carbohydrates, water and salts from plant.
   (b) • Glucose molecules in cellulose are link by chemical bond
       • that few enzymes can break down
       • no ruminants synthesise cellulose-splitting enzymes
       • anaerobic rumen bacteria break down cellulose
       • and release fatty acids, sugars, starch and vitamin B_{12}
       • which cow can absorb and utilise
       • in exchange bacteria are given a sheltered habitat
       • and obtain their inorganic nutrients from host.
   (c) • Bacteria break down organic wastes in faeces
       • and synthesise vitamin B and K
       • which are absorbed by the human body.
   (d) • Causes tuberculosis
       • damage the delicate lung tissue and breathing efficiency is reduced
       • irritation causes coughing and bacteria are spread

- blood vessels can burst and sputum may contain blood
- patient becomes exhausted and loses weight and may die.

9. (a) *Spirogyra*
   (b) Ribbon-like **chloroplast** is arranged spirally; to ensure absorption of maximum sunlight for photosynthesis.
   Many **pyrenoids** are present; for storage of starch as product of photosynthesis.
   Mucilage layer on outer surface of cell wall; traps oxygen gas to keep the plant afloat; in upper layers of water for effective interception of light.
   Large **vacuole**; contains osmotically active substances for absorption of water.
   Cell wall is permeable; water and CO_2 needed for photosynthesis enter freely.

2 Bryophytes and Pteridophytes

A. Bryophytes

Bryophytes are small, green plants. They include mosses, liverworts and hornworts. Examples of mosses are *Funaria* and *Polytrichum*.

1. Mosses are **terrestrial** plants. *Funaria* grows in cool, moist, shady places, like walls, rocks, along water furrows and margins of river-banks. It is fairly tolerant of desiccation.

2. In bryophytes the **gametophyte**, which consists initially of a branched, filamentous, **protonema**, is the **prominent phase** in the life cycle. The plant body does not contain highly specialised cells or tissues and is known as a **thallus** which consists of a simple "stem", "leaves" and **rhizoids**. The **sporophyte phase** is attached to the gametophyte and consists of a **foot**, **seta** and **capsule** (sporangium) in which spores are produced. The **calyptra**, which forms part of the gametophyte, protects the capsule.

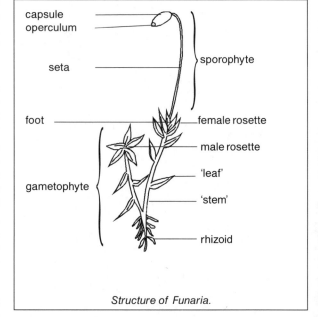

Structure of Funaria.

3. **The gametophyte phase** in mosses is green plants which contain chlorophyll and can photosynthesise. Water and mineral salts are absorbed by the **rhizoids**. Although the sporophyte contains chlorophyll it can not provide all the food it needs and is partially dependent on the gametophyte. In this respect it is **semi-parasitic**.

4. The life cycle shows an alternation between a haploid gamete-producing **gametophyte** and a diploid **sporophyte**, which produces haploid spores. This is called **alternation of generations**.
   The haploid gametophyte generation can only reproduce **sexually** and can only give rise to the sporophyte generation, while the diploid sporophyte generation can only reproduce **asexually** and give rise only to the gametophyte generation.
   The **antheridia** are the male sex organs which produce **antherozoids** (sperms). The **archegonia** are the female sex organs which produce **oospheres** (egg cells). The production of these gametes is by **mitosis**. Water is essential for fertilisation to take place. Sperms released from the antheridia swim to the archegonia and fusion results in a **diploid zygote**.
   Spore mother cells develop inside the sporangium and each undergoes **meiosis** forming haploid spores. In dry weather the spores are shedded from the sporangium and dispersed by wind. A spore germinates and gives rise to a **protonema**. Haploid moss plants develop from the buds on the protonema and the life cycle is completed.

5. The **ecological role** of bryophytes as **pioneers in succession**.

   They play a role in **plant succession**. Mosses colonise matlike layer of lichens and as **pioneer plants** they help to trap wind-blown soil and remains of dead leaves. In this way they help to increase the organic matter content and make the soil suitable for larger plants, such as ferns and even smaller animals. Bryophytes cover the soil surface and assist in binding the soil, retaining water and **reducing soil erosion**. They photosynthesise and assist in maintaining a favourable CO_2-O_2 balance in plant communities.

B. *Pteridophytes*

Pteridophytes include horse-tails, club-mosses and true ferns. Examples of ferns are *Dryopteris*, *Polystichum* and *Rumohra* (seven-week fern). Unlike bryophytes, pteridophytes have vascular tissue consisting of **xylem vessels** and **phloem tissue**.

1. Ferns are **terrestrial** plants. *Dryopteris* grows in moist, shady and woodland areas or gorges near streams in temperate regions.

2. In pteridophytes the **sporophyte** is the prominent phase with a long life-span. The sporophyte possesses roots, stems and leaves. The stem is a **rhizome**. Adventitious roots develop from the nodes of the rhizome. On the fronds develop **sori** with **sporangia**.

   The gametophyte is reduced to a small, simple **prothallus** without specialised tissue, and with a short life-span. The prothallus is a green, flat, heart-shaped structure about 10 mm long and 5 mm in diameter. On its lower surface it bears **rhizoids**, **archegonia** and **antheridia**.

3. The fully developed **sporophyte** in ferns is **autotrophic**. The root-hairs absorb water and dissolved mineral salts. Only during the early stage of its development the **embryo sporophyte** is dependent on the prothallus for its nutrition. The **prothallus** is completely **independent** and autotrophic.

4. In the life cycle of *Dryopteris*, like all pteridophytes, an **alternation of generations** occur. A **diploid** spore-producing generation, the **sporophyte** (*Dryopteris*-plant), alternates with a **haploid** gamete-producing generation, the **gametophyte** (prothallus).

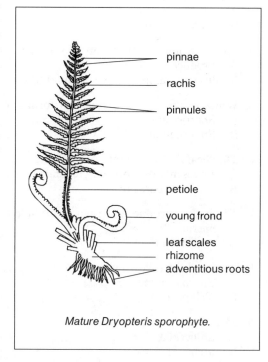

Mature Dryopteris sporophyte.

5. The **ecological role** of pteridophytes can be summarised as follows:
   (*a*) Play role as pioneers in **plant succession**. They form dense mats which trap wind-blown leaves and soil particles and create conditions for germination and growth of larger plants.

   (*b*) In areas which get less direct sunlight they are frequently found associated with tall grasses and shrubs. In such places it forms a **dense ground cover** and gives way to a **climax forest community**. In this way ferns form a lower stratum in a climax community and serve an important function in preventing soil erosion.

STANDARD GRADE
QUESTIONS

Section A

A. *Various possibilities are suggested as answers to the following questions. Indicate the correct answer.*

1. Which of the following plants are bryophytes?
   - A *Spirogyra*
   - B *Funaria*
   - C *Rhizopus*
   - D *Dryopteris*

2. Which of the following forms part of the sporophyte generation in bryophytes?
   - A Spore mother cells
   - B Protonema
   - C Adult moss plant
   - D Calyptra

3. Bryophytes are adapted to a terrestrial way of life, since they
   - A can photosynthesise.
   - B possess rhizoids.
   - C produce gametes.
   - D reproduce sexually.

4. The most important reason why *Funaria* is able to live in a terrestrial habitat, is because it
   - A produces gametes.
   - B lives in damp, shady surroundings.
   - C produces spores.
   - D is a thallus plant.

5. Which of the following does **not** form part of the sporophyte generation in bryophytes?
   - A Sporogonium
   - B Sporangium
   - C Calyptra
   - D Embryo

6. The "leafy" moss plant
   - A represents the sporophyte generation.
   - B is a semi-parasite.
   - C produces spores.
   - D is a thallus.

7. The male reproductive organ in bryophytes is the
   - A sperm.
   - B antheridium.
   - C sporangium.
   - D archegonium.

8. Which of the following does *not* form part of the gametophyte generation in bryophytes?
   - A Sporogonium
   - B Calyptra
   - C Paraphyses
   - D Antheridia

9. Meiosis occurs in *Funaria* in the
   - A antheridia.
   - B archegonia.
   - C antheridia and archegonia.
   - D capsule.

10. The gametophyte in bryophytes is the
    - A young embryo.
    - B adult plant.
    - C spores.
    - D prothallus.

11. Which of the following statement with regard to paraphyses in *Funaria* is **wrong**? Paraphyses
    - A occur among the sex organs.
    - B are multicellular.
    - C have diploid cells.
    - D are sterile.

12. The sporophyte in the moss plant is
    - A known as the protonema.
    - B the leafy stage.
    - C altogether independent.
    - D semi-parasitic.

13. The multicellular sterile organs which occur amongst the sex organs of bryophytes are known as
    A   paraphyses.   C   setae.
    B   antheridia.   D   sporangiophores.

14. Adventitious roots develop from
    A   tap roots.   C   radicles.
    B   stems.   D   lateral roots.

15. Binary fission is a method of reproduction that generally occurs in
    A   algae.   C   unicellular organisms.
    B   mosses.   D   fungi.

16. Which of the following is a product of fertilisation?
    A   Zygote   C   Antherozoid
    B   Spore   D   Gamete

17. In mosses gametes are produced by
    A   sporophytes.   C   antheridia and archegonia.
    B   zygotes.   D   sporangia.

18. Bryophytes are terrestrial plants that have no
    A   cellulose.   C   sporophyte phase.
    B   rhizoids.   D   vascular tissue.

19. The leafy fern plant
    A   represents the gametophyte generation.
    B   bears the sex organs.
    C   is a semi-parasite.
    D   has diploid cells.

20. A sorus is a collection of
    A   sporangia.   C   antheridia and archegonia.
    B   spores.   D   gametes.

21. The sporangium in *Dryopteris* ruptures because the
    A   indusium falls off.   C   placenta dries out
    B   stomium cells absorb water.   D   annulus cells shrink.

22. The prothallus of a fern develops
    A   from a germinating spore.   C   into an adult fern.
    B   roots, a stem and leaves.   D   after fertilisation has taken place.

23. This part of the life cycle in pteridophytes
    A   reproduces asexually.
    B   is part of the sporophyte.
    C   is dependent on the sporophyte.
    D   produces gametes.

24. In pteridophytes the sporophyte is
    A   capable of producing gametes.
    B   differentiated into roots, stems and leaves.
    C   much smaller than the gametophyte.
    D   known as the sporogonium.

25. The sporophyte of the fern plant
    A   has no roots.   C   possesses rhizoids.
    B   possesses a rhizome.   D   bears the reproductive organs.

26. In pteridophytes the process of meiosis takes place in the
    A   young frond.   C   sporangium.
    B   venter of the archegonium.   D   prothallus.

B. Write down the correct term for each of the following statements.

1. The structure that originates when the spore of a moss plant germinates
2. The gametophyte in ferns
3. The male reproductive organs of a moss plant
4. The female reproductive organs in pteridophytes
5. A group of sporangia usually beneath an indusium on the abaxial side of a frond
6. The elongated horizontal underground stem in pteridophytes
7. The phenomenon in some plants where both male and female sex cells are produced in the same individual
8. The name of the gametes which are produced in antheridia
9. The name of the female gamete which is produced in the venter of an archegonium in pteridophytes
10. The sterile structures which are found amongst the antheridia and archegonia in *Funaria*.
11. The fusion of a sperm and an egg to form a zygote
12. The basal, expanded portion of an archegonium which contains the oosphere
13. A membrane covering a sorus on a fern leaf
14. An old archegonial wall now forming a dry sheath covering the capsule of a moss sporophyte
15. A reproductive cell which develops into a new plant directly, without first fusing with another cell
16. A ring of thick-walled cells around the edge of a fern sporangium
17. The swollen tip of a moss sporophyte, supported by a slender stalk
18. An asexual reproductive cell.

C. Write down the letter of the description in column B which best suits the term in column A.

Column A		Column B
1. Prothallus	A	The male reproductive organs of mosses
2. Peristome	B	The group of plants to which ferns belong
3. Protonema	C	The structure in which spores are produced
4. Paraphysis	D	Ring of teeth-like projections around rim of mouth of sporangium in moss plant
5. Pteridophytes	E	The group of plants to which mosses belong
	F	A sterile filament in the reproductive apparatus of mosses
	G	A flattened, leaf-like structure resulting from the germination of a spore
	H	The first formed filament arising from the germinating spore of a moss

Section B

1. (*a*) Name the example of the moss plant you have studied.
   (*b*) Briefly describe the
      (i) natural habitat of the example mentioned, and
      (ii) structure of a "leaf" of a moss plant.

2. Study the following diagram of a moss plant and answer the questions.

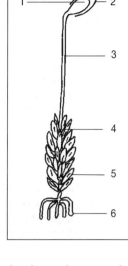

(a) Identify the structure labelled number 2. Is it haploid or diploid?
(b) What type of reproductive body is formed in the part numbered 1?
By which process is it formed?
(c) Identify structure 5.
Name its functions.
(d) Is the structure numbered 3, diploid or haploid?
(e) Identify structure 6.
Name its function.
(f) Write down the numbers of the part in the diagram which represent the sporophyte
(g) Give the name of the structure that develops when a spore of this plant germinates.
(h) Give the name of the organ in which the female gamete of this plant is formed.
(i) Draw a labelled diagram of the organ mentioned in (h).
(j) The structure numbered 5 has no cuticle. In what way does this limits the environment in which the moss can live?

3. Describe the nutrition of the
   (a) gametophyte, and
   (b) sporophyte in moss plants.

4. This diagram represents the tip of a female branch of a moss plant. Study it and answer the questions.

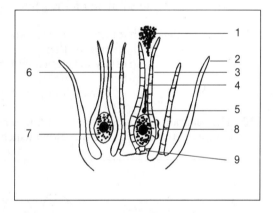

(a) To what group of plants does the moss plant belong?
(b) Identify the parts numbered 1 to 9.
(c) What is the male reproductive organ in mosses called? Draw a labelled diagram to illustrate its structure.
(d) Describe the process of fertilisation as it takes place in the moss plant.

5. Briefly describe the ecological role of bryophytes.

6. The diagram below represents a sorus in pteridophytes.

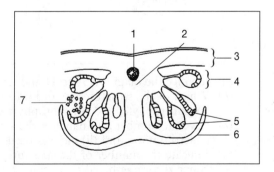

(a) Name the plant on which a sorus is found.
(b) Identify parts numbered 1 to 7.
(c) What is the function of part numbered (i) 3 and (ii) 6?
(d) Write down the number of a haploid structure.
(e) Draw a labelled diagram of the matured structure into which part numbered 7 grows after germination.

7. The diagram below represents part of a certain stage in the life cycle of a plant.
   (a) What is this structure/organ called?
   (b) On which plant is it found?
   (c) Identify the parts labelled 1 to 4.
   (d) With which process do you associate parts numbered 1 and 3?
   Briefly describe the process.

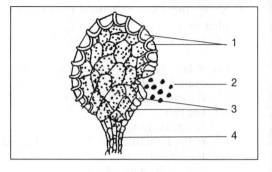

8. Study the diagram of a gametophyte.
   (a) What is this structure (organ) called?
   (b) Name the plant in whose life cycle this structure forms a part.
   (c) Is this structure haploid or diploid?
   (d) Identify the parts numbered 1 to 3.
   (e) Identify the parts numbered 4 and state its function.
   (f) Describe how fertilisation takes place in this structure.
   (g) Is this method of reproduction as described in (f), suitable for a successful survival on land? Give a reason for your answer.
   (h) There are no stomata on the gametophyte. Why is this not detrimental to this organism?

9. Describe the: (a) habitat of the fern; and
   (b) nutrition of the:
      (i) fully developed sporophyte;
      (ii) embryo sporophyte; and
      (iii) prothallus in pteridophytes.

10. The diagrams below represent certain structures in the life cycle of pteridophytes. Study the diagrams and answer the questions which follow.

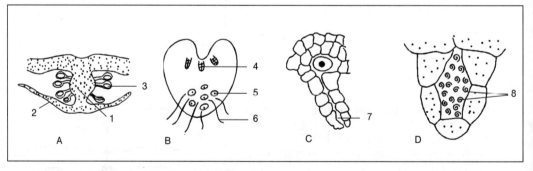

   (a) Identify the diagrams A to D and the parts numbered 1 to 8.
   (b) Which of the structures (A to D) form part of the
      (i) gametophyte, and (ii) sporophyte generation?
   (c) Give the number and name of the structure in which:
      (i) fertilisation, and (ii) meiosis occur.
   (d) Indicate the number of the structure in which spores are produced.
   (e) Name **three** numbered parts consisting of cells which are diploid.

11. Study the diagram of a young fern plant and answer the questions.
    (a) Identify the parts numbered 1 to 6.
    (b) Write down the number(s) of the part(s) that make(s) up the gametophyte.
    (c) Briefly describe the appearance of the part numbered 3.
    (d) Write down the number(s) of the part(s) that is/are diploid.
    (e) Identify the reproductive structures A and B. Briefly explain the significance of the position of these structures.

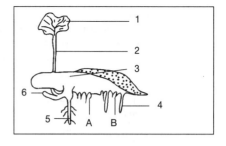

12. Briefly describe the ecological role of pteridophytes.

STANDARD GRADE
ANSWERS

Section A

A. 1. B   2. A   3. B   4. C   5. C   6. D   7. B
    8. A   9. D   10. B   11. C   12. D   13. A   14. B
    15. C   16. A   17. C   18. D   19. D   20. A   21. D
    22. A   23. D   24. B   25. B   26. C

B. 1. Protonema   2. Prothallus   3. Antheridia
    4. Archegonia   5. Sorus   6. Rhizome
    7. Monoecious   8. Antherozoids (sperms)   9. Oosphere (egg cell)
    10. Paraphyses   11. Fertilisation   12. Venter
    13. Indusium   14. Calyptra   15. Spore
    16. Annulus   17. Capsule/sporangium   18. Spore

C. 1. G   2. D   3. H   4. F   5. B

Section B

1. (a) *Funaria*
   (b) (i) • cool, damp and
           • shady surroundings
           • like walls, rocks, margins of freshwater rivers
           • fairly tolerant to desiccation.
       (ii) • yellowish-green in colour due to presence of chlorophyll
            • oval-like, pointed with smooth margin
            • sessile and spirally arranged on stem
            • blade consists of a single layer of cells
            • cuticle and stomata are absent.

2. (a) Calyptra; haploid   (b) Spore; meiosis
   (c) Leaf; produce food by photosynthesis; absorbs water

31

(d) Diploid
(e) Rhizoid; anchors plant to substrate; absorbs water and salts
(f) 1, 3, 4   (g) Protonema   (h) Archegonium
(i)

[Diagram of archegonium with labels: neck canal, neck cells (together forming neck); ovum; venter; stalk]

(j)
- Evaporation of water from leaf surface can not be prevented
- therefore, can only survive in damp, shady environment.

3. (a) *Gametophyte*
   - Is autotrophic
   - contains chlorophyll and can photosynthesise
   - raw materials for photosynthesis are absorbed from atmosphere
   - and soil, for synthesis of sugars
   - water and salts are absorbed by rhizoids
   - and water also by leaves.

   (b) *Sporophyte*
   - Foot of seta absorbs water and salts
   - from conducting tissue of gametophyte
   - lives semi-parasitic on gametophyte
   - therefore partially dependent on gametophyte for its nutrition
   - seta and base of sporangium
   - contain chlorophyll and stomata and can photosynthesise.

4. (a) Bryophytes   (b) 1 – Slimy glucose solution   2 – "Leaf"
   3 – Neck cell   4 – Neck canal   5 – Sperm   6 – Paraphysis
   7 – Egg cell/oosphere   8 – Venter   9 – Stalk
   (c) Antheridium

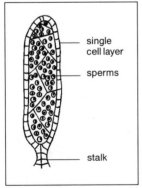

(d)
- Neck canal cells of archegonium
- secrete a slimy substance containing glucose
- antheridia burst and sperms/antherozoids are liberated
- and attracted by glucose secretion
- one swims down neck canal of archegonium
- fuses with egg/oosphere in venter
- to form a diploid zygote
- which marks the beginning of the sporophyte generation.

5. They cover soil surfaces and assist in binding soil and soil erosion is reduced.
Play a role in plant succession; mosses colonise mat-like layers of lichens.
As pioneers they trap wind-blown soil and remains of dead leaves in this way organic matter content is increased and make soil suitable for larger plants, e.g. ferns and even smaller animals to live and to establish themselves.
Since they photosynthesise a favourable CO_2-O_2 balance is maintained in plant communities.

6. (a) Fern
   (b) 1 – Vascular bundle/vein
       2 – Placenta
       3 – Pinna/leaf blade
       4 – Sporangium
       5 – Annulus cells
       6 – Indusium
       7 – Spores
   (c) (i) Protect sorus beneath;
           Photosynthesises and supplies food to sporangia
       (ii) Protects sporangia
   (d) No. 7
   (e) *Protallus of fern.*
       — apical notch
       — prothallus
       — archegonium
       — antheridium
       — rhizoid

7. (a) Sporangium (b) Fern
   (c) 1 – Annulus cells   2 – Spore   3 – Stomium cells   4 – Sporangiophore
   (d) Liberation of spores

   When the **spores ripen** the
   - indusium shrivels up and drops off
   - exposed sporangium dries out
   - thick-walled cells of the annulus shrink
   - tension is set up in each annulus cell
   - tension increases as whole annulus tries to straighten itself
   - thin-walled stomium cells eventually rapture and annulus curls back slowly
   - tension within each cell suddenly breaks
   - annulus snaps back to its original position slinging the spores into the air

8. (a) Prothallus   (b) Fern   (c) Haploid
   (d) 1 – Apical notch   2 – Archegonium   3 – Antheridium
   (e) Rhizoid – anchors prothallus in soil; absorbs water and salts
   (f) • Cap cell of antheridium is dislodged
       • antherozoids (sperms) are released and swim in thin film of water
       • neck canal cells of archegonium release malic acid which
       • stimulates movement of sperms and attracts them to archegonia
       • sperms swims upwards in neck canal
       • fuses with oosphere (egg cell) in venter
       • to form a diploid zygote.
   (g) Yes – The sunken archegonium gives greater protection to egg cell and embryo in the prothallus
   (h) Gametophyte is flat and thin; diffusion of gases takes place freely.

9. (a) Terrestrial – damp, shady, woodland areas
   (b) (i) *Fully developed sporophyte*
           Ferns are differentiated into roots, stems and leaves;
           leaf cells contain chloroplasts;
           and produce organic food by photosynthesis;
           as a result they are **autotrophic**.
           CO_2 diffuses through stomata to photosynthetic cells.
           Root-hairs absorb water and mineral salts;
           which are transported by conducting tissue to rest of plant.
           Rhizome contains stored food reserves;
           to be used during unfavourable growth periods,
       (ii) *Embryo sporophyte*
           Only during the early stages of its development;

the embryo sporophyte is **dependent** on gametophyte;
for its water, salts and organic nutrients.
Absorption takes place through foot;
which is embedded in prothallus.
Embryo forms its own roots and leaves with chloroplasts
and becomes autotrophic and completely **independent**.
(iii) *Prothallus*
Is completely **independent** and autotrophic;
water and salts are absorbed from soil by rhizoids.

10. (*a*) A – Sorus   B – Prothallus   C – Archegonium   D – Antheridium
    1 – Placenta   2 – Indusium   3 – Sporangium   4 – Archegonium
    5 – Antheridium   6 – Rhizoid   7 – Neck canal   8 – Sperms
    (*b*) (i) B, C, D (ii) A
    (*c*) (i) 4 – Archegonium   (ii) 3 – Sporangium
    (*d*) No. 3 – sporangium   (*e*) No.1, 2 and 3.

11. (*a*) 1 – first leaf of sporophyte 2 – Stem 3 – Prothallus tissue
    4 – Rhizoid 5 – Primary root 6 – rhizome
    (*b*) 3; 4   (*c*) Flat, green, heart-shaped   (*d*) 1; 2; 5; 6
    (*e*) A – Archegonium   B – Antheridium
    • Being on ventral side of prothallus they enjoy protection
    • Prothallus conserves film of water in which sperms swim.

12. Play role as pioneers in plant succession;
    forms dense mats which trap;
    wind-blown leaves and soil particles;
    create conditions for germination and growth of larger plants.
    In areas where they receive less direct sunlight;
    they form dense ground cover;
    gives way to a climax forest community;
    and serves a function in preventing soil erosion.

HIGHER GRADE
QUESTIONS

Section A

A. Various possibilities are suggested as answers to the following questions. Indicate the correct answer.

1. Which statement regarding the moss plant is *wrong*?
    A   The sporophyte generation is mainly heterotrophic
    B   The gametophyte has no true roots but instead multicellular rhizoids
    C   Heterogametes fuse to give rise to the protonema
    D   Spore mother cells develop inside the capsule

2. The sporogonium of *funaria* consist of a
    A   sporangium, stalk and foot.
    B   stem, rhizoids and leaf-like structures.
    C   sporangium, seta and rhizoids
    D   foot, sporangium and leaf-like structures

Questions 3 to 5 refer to the following diagram.

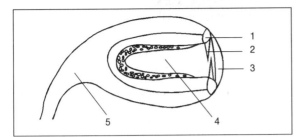

3. In the diagram 1 refers to
   A   annulus.
   B   row of peristome teeth.
   C   columella.
   D   operculum.

4. The part where photosynthesis takes place, is indicated by
   A   2.
   B   3.
   C   4.
   D   5.

5. The part which is hygroscopic and involved in the release of spores, is indicated by
   A   1.
   B   2.
   C   3.
   D   4.

Questions 6 to 9 refer to the diagram of a moss plant.

6. In this diagram all parts of the gametophyte generation are represented by numbers
   A   1, 5, 6 and 7.
   B   4, 5, 6 and 7.
   C   1, 2, 3 and 4.
   D   1, 2 and 3.

7. The phase in the life cycle of a moss plant which is comparable to that of the fronds, rhizome and roots of a fern, is represented by numbers
   A   4, 5 and 7.
   B   2, 3 and 4.
   C   4, 5 and 6.
   D   5, 6 and 7.

8. The parts best suited to a moist habitat are represented by numbers
   A   1, 2 and 3.
   B   4, 6 and 7.
   C   5, 6 and 7.
   D   4, 5 and 6.

9. The small size of a moss plant is a consequence of rudimentary development of certain structures, which are represented in this diagram by numbers
   A   2 and 3.
   B   6 and 7.
   C   1 and 2.
   D   3 and 4.

10. In the alternation of generations, a zygote is
    A   diploid.
    B   part of the gametophyte generation.
    C   sunken into the sporangium.
    D   a product of spore germination.

11. If a **spore** in the structure numbered 1 has **ten** chromosomes in its nucleus, how many chromosomes will there be in the nucleus of a cell making up the part numbered 2?

    A   5
    B   10
    C   20
    D   40

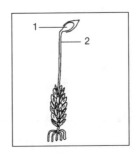

12. The result of meiosis occuring in a single spore mother cell, is
    A 2 diploid and 2 haploid cells.
    B 4 diploid cells.
    C 2 diploid cells.
    D 4 haploid cells.

13. Meiosis in plant species occurs in
    A each generation.
    B the reproductive organs.
    C every other generation.
    D the gametophytes.

14. If a cell of the sporangium of a fern possesses 20 chromosomes in the nucleus, 10 chromosomes will occur in a cell of the
    A sorus.
    B antheridia.
    C adventitious roots.
    D young frond of fern.

15. In the example of bryophytes the gametophyte generation is
    A the adult plant.
    B supplied with root hairs.
    C heterotrophic.
    D dependent on the sporophyte generation for nutrition.

16. Which of the following forms part of the sporophyte generation in bryophytes?
    A Adult moss plant
    B Spores
    C Spore mother cells
    D Calyptra

17. Which of the following chromosome numbers is stated incorrectly?
    (n = haploid)
    A Oosphere of *Dryopteris*   .... n
    B Prothallus of pteridophytes 2n
    C Spore of bryophytes   .... n
    D Zygospore of *Spirogyra* 2n

18. Bryophytes are terrestrial plants that have no
    A vascular tissue.
    B leaf-like structures.
    C water absorption structures.
    D cellulose.

19. A spore-producing plant whose stem is underground, and whose leaves are long and divided into many leaflets, is probably a
    A conifer.
    B moss.
    C flowering plant
    D fern.

20. Rhizoids serve a function in mosses that is served in most other terrestrial plants by
    A leaves.
    B stomata.
    C roots.
    D stems.

Questions 21 to 25 refer to the following diagram which represents the life cycle of a moss plant.

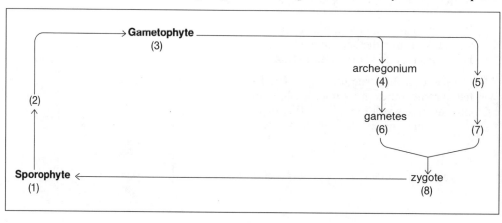

21. An oosphere is indicated by number
    A 2.
    B 5.
    C 6.
    D 7.

22. Meiosis occurs between number
    A 1 and 2.
    B 3 and 4.
    C 5 and 7.
    D 6 and 8.

23. A sporogonium is indicated by number
    A 1.
    B 2.
    C 5.
    D 8.

24. Number 2 indicates
    A an embryo sporophyte.
    B spores.
    C a structure with 2n number of chromosomes
    D antherozoids.

25. The dominant phase is indicated by number
    A 1.
    B 3.
    C 5.
    D 7.

26. The embryo and the very young fern plant
    A are independent.
    B live saprophytically.
    C obtain food from the gametophyte.
    D are autotrophic.

27. In the alternation of generations, the haploid phase is the
    A sporogonium.
    B sporophyte.
    C sporangium.
    D gametophyte.

28. If a plant is homosporous it will produce
    A one type of gametophyte.
    B diploid spores.
    C two kinds of spores.
    D sperms cells only.

29. Leaves that bear sporangia, are called
    A sori.
    B cones.
    C sporogonia.
    D sporophylls.

30. There are 24 chromosomes in each spore in pteridophytes. How many chromosomes are there in the nucleus of a prothallus cell?
    A 12
    B 24
    C 48
    D 2n

Questions 31 to 34 refer to the following diagram which represents a typical life cycle of a sexually reproducing plant.

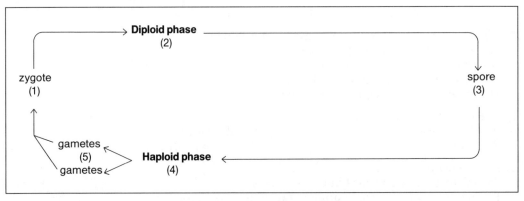

31. Meiosis occurs between
    A 2 and 3.
    B 3 and 4.
    C 4 and 5.
    D 5 and 6.

37

32. If the life cycle was that of a fern, the dominant phase would be number
    A 1.
    B 2.
    C 3.
    D 4.

33. In most fresh water algae the phase usually present during favourable growth conditions, is number
    A 2.
    B 3.
    C 4.
    D 5.

34. The calyptra on a moss sporangium originally formed part of
    A 1.
    B 2.
    C 3.
    D 4.

35. Which of the following have fungi, algae, mosses and ferns in common? They all
    A shows alternation of generations.
    B possess photosynthetic tissue.
    C produce spores for dispersal.
    D reproduce sexually by conjugation.

Questions 36 to 38 refer to the following diagrams.

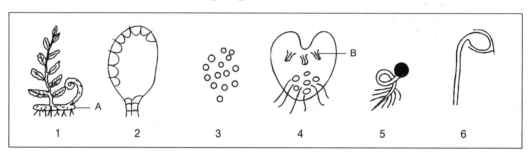

36. Which of the diagrams in the life cycle of a pteridophyte-species form part of the sporophyte generation?
    A 1 and 2
    B 4 and 5
    C 3 and 6
    D 3, 4 and 5

37. A refers to
    A a root.
    B a tuber.
    C a stolon.
    D an underground stem.

38. B refers to
    A a sorus.
    B an antherozoid.
    C an archegonium.
    D a rhizoid.

39. What happens to the fertilised oosphere of a fern? It
    A grows into an asexual protonema.
    B grows into a sexual prothallus.
    C becomes a spore which rapidly dies.
    D grows into a spore-producing plant.

40. The prothallus of the fern is a
    A sporophyte.
    B gametophyte.
    C dependent plant.
    D spore producer.

41. In the life-cycle of a fern
    A microspores are formed in microsporangia.
    B zygospores are formed during unfavourable conditions.
    C zygotes are formed within antheridia.
    D zygotes are formed within archegonia.

B. Write down the correct term for each of the following statements.

1. The generation in the life cycle of a moss where all the cells are diploid
2. The diploid cell in bryophytes which gives rise to spores during the process of meiosis
3. The brown scale leaves that protect the young developing frond of the fern
4. The row of cells in the wall of the sporangium in *Polystichum* or *Dryopteris* which is concerned with the liberation and dispersal of spores
5. The basal, expanded portion of an archegonium in a fern gametophyte which contains an oosphere
6. The phase in the life cycle of pteridophytes which produces gametes
7. A chromosome number, the n number, characteristic of gametes
8. In plants, a life cycle in which the haploid gametophyte produces gametes that upon fusion form a zygote that develops into a diploid sporophyte
9. An asexual reproductive cell that is capable of developing into an adult organism without fertilisation
10. Having the reproductive organs in separate structures, but borne on the same plant, as in *Funaria*

C. The following is a schematic representation of the life cycle of a fern.

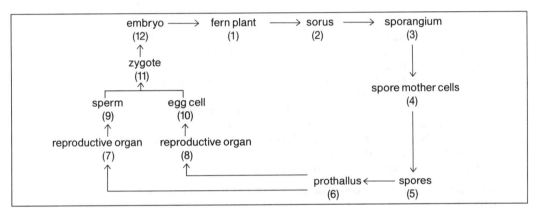

Complete the following questions and use relevant numbers where required.
(a) Meiosis occurs between No. __ and No. __
(b) The sporophyte generation is represented by Nos. __
(c) The reproductive organ No. 7 is the ____ and No. 8 the ____
(d) The type of cell division taking place between No. 11 and No. 12 is ____
(e) Draw a line across the diagram to separate the diploid and haploid generations.

Section B.

1. This diagram represents a longitudinal section through the capsule of a moss plant.

   (a) Identify the parts numbered 1 to 10.
   (b) List the numbers and names of all those parts concerned with supplying nutrition to the capsule.
   (c) Which numbered part is haploid?
   (d) Briefly describe the function of part numbered 2.

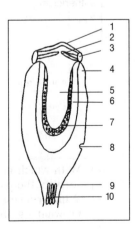

2. The diagram below represents the life cycle in bryophytes.

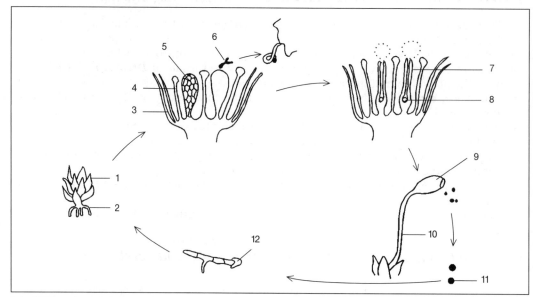

(a) Write down the numbers and names of two different structures which photosynthesise.
(b) Name the structure numbered 2 and state its function.
(c) Describe the role played by the parts numbered 3 to 8 in sexual reproduction.
(d) Write down the numbers of the parts which represent the sporophyte generation.
(e) Name the structure numbered 11.
What is the chromosome composition of this structure?
By what process is structure No. 11 produced?
(f) What is the main difference in structure between structure 2 and root-hairs?
(g) Identify the parts numbered 9, 10 and 12.
(h) Describe fertilisation in *Polytrichum* or *Funaria*.

3. Mention the name of a plant where each of the following occurs, and state one function of each:
(a) pyrenoid   (b) protonema   (c) ribbon-like chloroplast
(d) ramenta   (e) dwarf shoots   (f) rhizome
(g) mycelium   (h) rhizoid

4. Stydy the schematic representation of the life cycle in a type of plant you have studied and answer the questions.

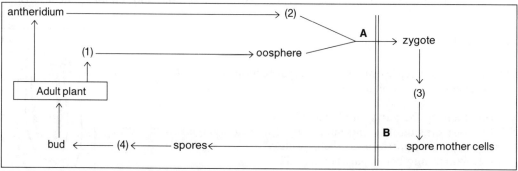

(a) With which group of plant, and specific example, is this diagram associated?
(b) Identify the parts numbered 1 to 4.
(c) Which process takes place at (i) A, and (ii) B?
Of what significance is the process, named in (ii), to the plant concerned?

(d) Compare the gametophyte generation of the example named in (a), to a specific example of pteridophytes you have studied with regard to:
   (i) nutrition (two similarities);
   (ii) transport of the male gametes (two similarities and two differences); and
   (iii) structure of female sex organs (two differences).

5. The diagram below represents the life cycle of a fern-plant.

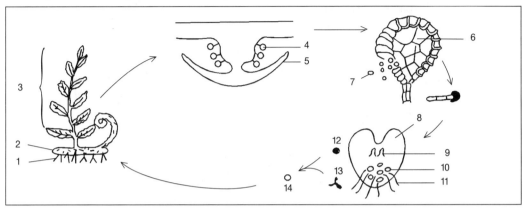

(a) State the name of the fern you have studied.
(b) Write down the numbers and names of **two** structures in this life cycle which can photosynthesise.
(c) Which **two** of the following numbered parts have haploid nuclei?
   2; 5; 7; 11
(d) Which **two** of the following numbered parts have diploid nuclei?
   6; 7; 10; 14
(e) (i) Identify the stucture numbered 7.
    (ii) By what process is this structure produced?
    (iii) What is the significance of this process with reference to reproduction and alternation of generations?
(f) Which generation is structurally best suited to a successful life on land?
   Give **two** reasons to support your answer.
(g) Identify the structure numbered 8.
   What is the chromosome composition of this structure?
(h) Indicate the number of the part where:
    (i) slimy malic acid solution is secreted;
    (ii) circinate vernation occurs; and
    (iii) antherozoids are produced.

6. (a) Briefly explain what is meant by alternation of generations.
   (b) Draw a labelled diagram to show the structure of a mature
       (i) female organ on the gametophyte of a moss plant; and
       (ii) male organ on the gametophyte of a fern plant.

7. Study the diagram of a sorus of a fern and answer the questions.
   (a) Of which generation does the sorus form part?
   (b) Exactly where is a sorus found on the fern?
   (c) Identify the parts numbered 1 to 6.
   (d) What is the function of the part numbered 6?
   (e) What process takes place in the part numbered 2, and why?
   (f) With which process do you associate the part numbered 5?
       With which structures in bryophytes could you associate this process?

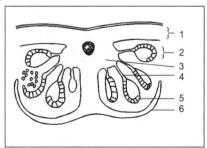

(g) Is the sorus in this diagram ripe? State **two** reasons.
(h) What develops from the spore?

8. The following diagram represents a certain structure in the life cycle of a fern plant.
   (a) To which group of plants does the fern belong?
   (b) Which generation is represented by (i) A, and (ii) B?
   (c) Identify the parts numbered 1 to 5.
   (d) What eventually happens to each of the parts mentioned in (c)?
   (e) In which respect does the nutrition of this generation correspond with the same generation in bryophytes?
   (f) If a cell in part numbered 4 possesses 10 chromosomes in the nucleus, what will the number of chromosomes be in the nucleus of a cell in part numbered (i) 2, (ii) 3 and (iii) 5?
   (g) State **three** similarities and **one** difference between the generation of ferns and mosses as indicated by A.

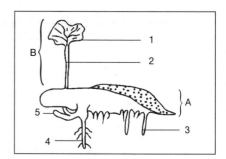

Section C.

1. 'The sporophyte and gametophyte generations in the life cycle of plants can, in general, be regarded as phases suited to terrestrial and aquatic conditions'.
   Explain this statement referring to the life cycles of an example of a bryophyte and an example of a pteridophyte.

2. Describe in what ways a named example of each of the following group of plants in the reproduction processess is dependent on water and how each is adapted to cope with a water shortage:
   (a) **mycophyte species;**
   (b) **phycophyte species;** and
   (c) **bryophyte species.**

3. In the life cycle of bryophytes and pteridophytes an alternation of generations occurs. The simplified diagrams below represent the alternation of generations of a moss plant and a fern. The nucleus in the leaf of a moss plant contains **10** chromosomes and that of a frond **20**. Study the diagrams and answer the questions.

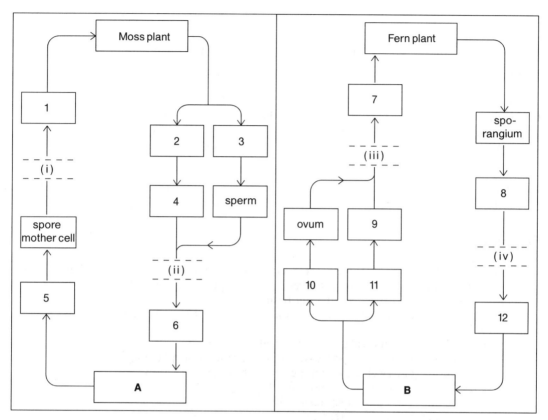

(a) What is meant by the term alternation of generations?
(b) Identify the processes numbered (i) to (iv).
(c) (i) Name the moss plant and fern that you have studied.
    (ii) What part of the life cycle do A and B represent respectively?
(d) Answer the following questions concerning structure A.
    (i) What is the name of this structure?
    (ii) State the chromosome number of this structure.
    (iii) Briefly describe the nutrition of this structure.
    (iv) Draw a labelled diagram of this structure.
(e) (i) Identify the structures numbered 1 to 6.
    (ii) State the chromosome number of structures 1, 4 and 6 respectively.
    (iii) Where are structures numbered 2 and 3 found on the moss plant respectively?
(f) Briefly describe where and how process (ii) takes place.
(g) What is the name of structure B? Briefly describe its structure.
(h) (i) Identify the structures numbered 7 to 11.
    (ii) State the chromosome number of structures 7, 8, 9 and 12 respectively.
(i) Briefly describe the process in which structure 12 is released.

HIGHER GRADE
ANSWERS

Section A.

A. 1. C   2. A   3. A   4. D   5. B   6. A   7. D
    8. C   9. B   10. A 11. C 12. D 13. C 14. B
    15. A 16. C 17. B 18. A 19. D 20. C 21. C
    22. A 23. A 24. B 25. B 26. C 27. D 28. A
    29. D 30. B 31. A 32. B 33. C 34. D 35. A
    36. A 37. D 38. C 39. D 40. B 41. D

B. 1. Sporophyte   2. Spore mother cell   3. Ramenta
    4. Annulus   5. Venter   6. Gametophyte
    7. Haploid   8. Alternation of generations 9. Spore
    10. Monoecious

C. (*a*) 4; 5   (*b*) 1, 2, 3, 4, 11, 12   (*c*) antheridium; archegonium
    (*d*) mitosis   (*e*) draw line from spores to zygote, i.e. 5 to 11

Section B.

1. (*a*) 1 – operculum   2 – peristome teeth   3 – annulus   4 – photosynthetic tissue
          5 – columella   6 – air chamber   7 – spore   8 – stoma
          9 – seta   10 – conducting tissue
   (*b*) 4 – photosynthetic tissue   8 – stoma   10 – conducting tissue
   (*c*) No. 7
   (*d*) • Peristome teeth concerned with dispersal of spores
          • they are hygroscopic, i.e. sensitive to moisture in atmosphere
          • when moist they swell and opening remains closed
          • when dry they pull away from each other
          • the opening enlarges and spores easily fall out
          • to be dispersed by air currents.

2. (*a*) 1 – leaf ; 9 – sporangium
   (*b*) rhizoid – anchors plant; absorbs water and mineral salts
   (*c*) 3 – leaves of male rosette; protect sex organs
          4 – paraphysis; secretes liquid which prevents sex organs from drying out
          5 – antheridium; produces antherozoids (sperms)
          6 – antherozoid; necessary for fertilisation process
          7 – archegonium; produces oosphere (egg)
          8 – oosphere; which develops into sporophyte after fertilisation
   (*d*) 9 and 10
   (*e*) 11 – spore; haploid; meiosis
   (*f*) rhizoid – multicellular; root hair – unicellular
   (*g*) 9 – sporangium   10 – seta   12 – protonema
   (*h*) • Ripe antheridium bursts at top under wet conditions
          • antherozoids escape and swim in thin film of water
          • with aid of two whip-like flagella
          • neck cells and ventral canal cell of archegonium dissolve
          • to form passage for antherozoids
          • neck canal of archegonium secretes slimy glucose fluid
          • which attracts antherozoids (chemotaxis)
          • one antherozoid swims down neck canal
          • fuses with oosphere in venter to form diploid zygote

3. (a) **Pyrenoid:** *Spirogyra* – stores starch
   (b) **Protonema:** *Funaria* – develops into new gametophyte generation
   (c) **Ribbon-like chloroplast:** *Spirogyra* – produces organic food by process of photosynthesis
   (d) **Ramenta:** *Dryopteris/Polystichum* – protect developing fronds against desiccation
   (e) **Dwarf shoots:** *Pinus* – bear needle-like leaves
   (f) **Rhizome:** *Dryopteris* – forms nodes from which leaves and roots arise
   (g) **Mycelium:** *Rhizopus* – forms stolons, rhizoids and sporangiophores, each with its own function
   (h) **Rhizoid:** *Funaria/Polytrichum* – anchors plant and absorbes water and mineral salts

4. (a) Bryophytes – *Polytrichum/Funaria*
   (b) 1 – archegonium 2 – antherozoid 3 – sporangium 4 – protonema
   (c) (i) fertilisation (ii) meiosis – ensures that number of chromosomes in cell of successive generations will remain constant
   (d) (i) • independent autotrophic green plant
       • rhizoids absorb water and mineral salts
       (ii) Similarities • swim; in film of watery solution
            Differences: **bryophytes** • swim with aid of **two** whip-like flagella
                         • in slimy **glucose** solution
                         pteridophytes • swim with aid of **several** flagella
                         • in slimy **malic acid** solution
       (iii) Differences: **bryophytes** • archegonium possesses **long** straight neck
                          • attached to gametophyte by means of a short **stalk**
                          pteridophytes • archegonium prossesses **short** bended neck
                          • **embedded** into cells of prothallus

5. (a) Dryopteris   (b) 3 – frond ; 8 – prothallus   (c) 7 ; 11
   (b) 6 ; 14   (e) (i) spore (ii) meiosis (iii) by meiosis the diploid sporophyte give rise to haploid gametophyte;
   when spore mother cells form spores;
   gametophyte and sporophyte alternate regularly with one another
   (f) sporophyte • is differentiated into roots, stems and leaves
                  • conducting and strengthening tissues are present
   (g) prothallus • haploid   (h) (i) No. 9 (ii) No. 3 (iii) No. 10

6. (a) • That is a sexual life cycle where a gamete producing generation
       • with a single set of chromosomes, haploid (n), in each nucleus
       • the gametophyte
       • alternate regularly with a spore producing generation
       • with a double set of chromosomes, diploid (2n), in each nucleus
       • the sporophyte.
   (b)

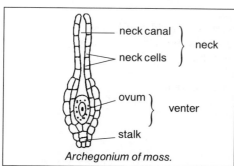
Archegonium of moss.

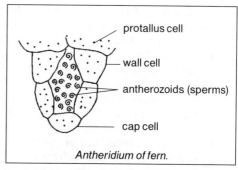
Antheridium of fern.

7. (a) Sporophyte   (b) on abaxial side of frond; along a vein; attached to placenta
   (c) 1 – leaf blade   2 – sporangium   3 – placenta
       4 – sporangiophore   5 – annulus cells   6 – indusium

- (d) protects sporangia against desiccation / injury
- (e) meiosis – to produce haploid spores / restore haploid state
- (f) liberation and dispersal of spores – peristome teeth
- (g) yes • sporangium has burst open
    - • indusium has pulled away from epidermis and starts to shrivel up
- (h) prothallus

8. (a) Pteridophytes
   (b) A – gametophyte (prothallus);   B – sporophyte (embryo sporophyte)
   (c) 1 – primary leaf   2 – stem (**not** rhizome)   3 – rhizoid
       4 – primary root   5 – rhizome
   (d) Nos. 1, 2, 3 and 4 wither and die
       No. 5 produces leaves and adventitious roots of developing fern
   (e) both are autotrophic and synthesise own organic food
   (f) (i) 10 (ii) 5 (iii) 10
   (g) **Similarities:** • reproductive organs are antheridia and archegonia
       - • they are autotrophic and independent
       - • are thalli
       Difference: mosses – gametophyte is conspicuous and prominent generation
       ferns – inconspicuous, much reduced and short-lived

Section C.

1. **Sporohyte suited to terrestial conditions**
   Bryophyte: **moss** – *Polytrichum* or *Funaria*
   - sporogonium is anchored by foot in apex of gametophyte
   - can withstand the buffering effects of wind
   - outer epidermal cells of capsule is hardened and thickened
   - to support sporophyte in air
   - calyptra protects capsule
   - stomata are present for gaseous exchange
   - sporogenous tissue is deeply situated, thus well protected
   - capsule dries out when spores are ready for dispersal
   - calyptra and operculum fall off
   - exposing peristome teeth to dry air
   - in damp weather they remain close
   - when dry peristome teeth open outwards and spores are exposed
   - long seta holds capsule well into air
   - and permits capsule to sway and scatter spores in light breeze

   Pteridophyte: **fern** – *Dryopteris* or *Polystichum*
   - rhizome covered with tough epidermis, ramenta and old leaf bases
   - in this way rhizome is protected from excessive water loss
   - rhizome stores food to be used during unsuitable growth conditions
   - **fronds** well supported in air by tough rachis
   - closely packed upper and lower epidermis with cuticle
   - form effective barrier and reducing water loss
   - stomata on abaxial side for gaseous exchange
   - circinate vernation in young frond
   - provides protection of growing point
   - **spores** develop inside sporangium
   - **sporangium** is on lower surface of frond
   - and protected by indusium
   - structure of sporangium adapted to spore dispersal in dry conditions
   - annulus cells begin to dry and tension is set up

- causing thin-walled stomium cells to tear
- annulus curls back, tension breaks and annulus snaps back
- spores are slung into the air and dispersed by wind

Gametophyte suited to aquatic conditions

Moss plant
- stems and leaves lack a well differentiated conducting tissue
- leaves are only one cell thick and stomata are absent
- cuticle is absent and gaseous exchange through thin wall
- leaves absorb water
- water is necessary for **fertilisation** to take place
- biflagellate antherozoids swim to archegonia
- attracted by glucose solution secreted at neck opening
- since gametophyte plant is short it is easily covered by water / dew

Fern plant
- prothallus only develops in moist surroundings
- fertilisation is dependent on water
- antherozoids possess several flagella
- and swim from antheridium to archegonium
- attracted by malic acid secretion
- flattened form permits it to be easily covered by water

2. (a) **Mycophyte:** *Rhizopus* sp.

dependent
- when growth conditions are moist asexual reproduction occurs
- in sporangium many spores are produced
- columella bulges inwards
- sporangium absorbs water to soften wall
- columella causes pressure on sporangium wall which bursts and spores are released

adaptation
- when surrounding becomes dry sexual reproduction occurs
- gametes fuse by conjugation and zygote forms
- zygote rounds off forming a zygospore
- zygospore enlarges and thick, hard outer walls form
- in dormant state it can withstand long dry conditions

(b) **Phycophyte:** *Spirogyra* sp.

dependent
- when there is sufficient water asexual reproduction occurs
- filament divides mitotically
- each fragment of filament forms a new filament
- this is called fragmentation

adaptation
- when surrounding water dries up sexual reproduction occurs
- which may be scalariform or lateral conjugation
- protoplasm of gametes conjugates
- zygote develops into zygospore
- which develops a hard, waterproof, resistant capsule
- can withstand unfavourable dry environmental conditions

(c) **Bryophyte:** *Polytrichum* sp. / *Funaria* sp.

dependent
- in damp soil the exine of spore absorbs water
- and intine grows out to form protoplasm

- mature antheridia absorb water
- when they ripen they at top under moist conditions
- antherozoids escape and swim with aid of flagella
- in moist conditions neck cells of archegonia dissolve
- to form passage for antherozoids
- tip of neck canal secretes slimy glucose liquid
- antherozoids are attracted to archegonia by this substance
- along neck canal to oosphere in venter

adaptation
- as spores become mature the capsule becomes dried out
- calyptra and operculum fall off
- to expose peristome teeth to dry air
- peristome bends outwards and stomiun opens when air is dry
- when capsule sways on its long thin seta
- spores are scattered under dry conditions and dispersed by wind
- spores are protected by tough exine against desiccation

3. (a) • it is the alternation in the life cycle between
 • a haploid gamete-producing gametophyte and
 • a diploid spore-producing sporophyte
 • gametophyte produces sexually and the sporophyte asexually
 (b) (i) meiosis (ii) fertilisation (iii) fertilisation (iv) meiosis
 (c) (i) Moss – *Polytrichum;* Fern – *Polystichum*
   (ii) A – sporophyte; B – gametophyte
 (d) (i) Sporogonium (ii) 20
   (iii) • Foot of seta absorbs water and salts
    • from conducting tissue of gametophyte
    • lives semi-parasitic on gametophyte
    • therefore partially dependent on gametophyte for its nutrition
    • seta and base of sporangium
    • contain chlorophyll and stomata and can photosynthesise.
   (iv) **Labels:** sporangium, operculum, seta, foot
 (e) (i) 1 – spore 2 – archegonium 3 – antheridium
   4 – ovum 5 – sporangium 6 – zygote
   (ii) 1 – 10 4 – 10 6 – 20
   (iii) 2 • on wider female rosette between paraphyses
    • at tip of female shoot
    • in midst of cluster brown-coloured leaves
   3 • on narrow male rosette between paraphyses
    • at tip of another shoot
    • in midst of cluster green leaves
 (f) • Ripe antheridium bursts at top under wet conditions
  • antherozoids escape and swim in thin film of water
  • with aid of two whip-like flagella
  • neck cells and ventral canal cell of archegonium dissolve
  • to form passage for antherozoids
  • neck canal of archegonium secretes slimy glucose fluid
  • which attracts antherozoids (chemotaxis)
  • one antherozoid swims down neck canal
  • fuses with oosphere in venter to form diploid zygote.
 (g) Prothallus • green, flat, thin and heart-shaped
    • rhizoids, archegonia and antheridia occur on ventral side
 (h) (i) 7 – zygote 8 – spore mother cell 9 – sperm
   10 – archegonium 11 – antheridium
   (ii) 7 – 20 8 – 20 9 – 10 12 – 10

(*i*) No. 12 is a spore
- when the spore is ripe
- indusium shrivels up and drops off
- exposed sporangium dries out
- thick-walled cells of the annulus shrink
- tension is set up in each annulus cell
- tension increases as whole annulus tries to straighten itself
- thin-walled stomium cells eventually rapture and annulus curls back slowly
- tension within each cell suddenly breaks
- annulus snaps back to its original position slinging the spores into the air

3 Spermatophytes – Gymnosperms and Angiosperms

A. General

1. **Spermatophytes** are "seed plants" and include **gymnosperms**, which produce seeds which are not enclosed in an ovary wall, and **angiosperms**, which produce flowers and fruit with seeds enclosed in an ovary wall. A **seed** is a dormant embryonic diploid **sporophyte plant** which is completely independent of the parent plant.

2. Spermatophytes contain **chloroplasts** and synthesise their own organic food and are **autotrophic**. The sporophyte has roots which absorb water and salt ions. The stems have conducting tissue for transport of water, salts and produced food. The haploid gametophyte is **dependent** on diploid sporophyte for its nutrition.

3. In the life cycle of Spermatophytes there is an **alternation of generations**. These plants continue the tendency seen in Pteridophytes to **reduce the gametophyte** which is incorporated into the body of the sporophyte. The male gametophyte is represented by the content of the **pollen grain** and **pollen tube** and the female gametophyte by the content of the **embryo sac**. The female gametophytes are attached to and protected by the sporophyte.

    The rise in size of the sporophyte and decline in size of the gametophyte can be seen as an **adaptation to life** in a terrestrial environment.

4. The process by which pollen is transferred from the male parts to the female parts is known as **pollination**. In Spermatophytes the sporophyte produces two types of spores: **microspores** which give rise to the male gametophytes and **megaspores** which develop into the female gametophytes; we say Spermatophytes are **heterosporous**. **Fertilisation** is achieved by the formation and growth of a **pollen tube** which conveys the male nuclei to those of the female in the female gametophyte. After fertilisation the zygote develops into an **embryo**. The production of pollen and pollen tubes freed spermatophytes from dependence on free water for fertilisation.

B. Gymnosperms

Gymnosperms (Gr. **gymnos**: naked; **sperma**: seed) include cycads and conifers, e.g. *Pinus* spp. which bear cones.

1. *Pinus* spp. are exotic and **terrestrial**. They grow in infertile, sandy soil in areas where it is dry, cold and windy.

2. *Pinus* spp. have needle-like leaves, well developed stems and a tap root system with lateral roots. The pine is **monoecious** and bears male and female **cones** on the same tree. The **male cones** occur in the **place of dwarf shoots** and each consists of a central axis around which **microsporophylls** (scale leaves) are spirally arranged, each with two **microsporangia** (pollen sacs) in which **pollen grains** are produced. The **female cones** occur in the **place of long shoots** and each consists of **megasporophylls** (ovuliferous scales) each carries two **megasporangia** (ovules) on its upper surface. The ovule is enclosed by an **integument** and **nucellus**.

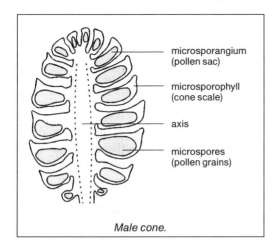

Male cone.

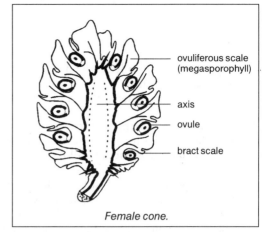

Female cone.

3. Male cones produce **microspores** (pollen grains) and female cones megaspores after meiosis has taken place. Pollen grains are dispersed by wind and enter the **micropyles** of female ovules. A megaspore develops into a female gametophyte with **archegonia** each containing an egg cell.

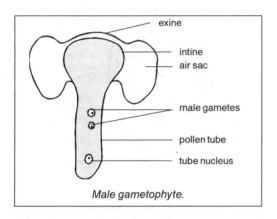

Male gametophyte.

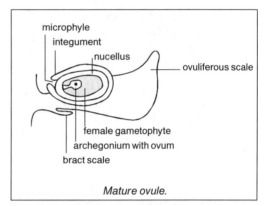

Mature ovule.

4. A pollen grain grows a **pollen tube** which enters the archegonium and releases a male gamete which fuses with the egg cell producing a diploid **zygote**. The zygote develops in an **embryo** surrounded by **endosperm** and **testa** (seed coat). The ovule is now called a **seed** and contains a radicle, plumule and several cotyledons. When conditions become favourable for growth, the seed germinates to produce a seedling which grows into a mature sporophyte tree.

5. The **ecological role** of Gymnosperms can be summarised as follows:
   (a) **Effect of plants on soil.** Fallen pine leaves decompose slowly and humic acids accumulate with the result that few humus is formed and the soil becomes infertile.
   (b) **Mycorrhiza.** A symbiotic relationship between the pine roots and a fungal mycelium exists by which water and dissolved mineral salts from the soil are transported by the hyphae of the fungus to the cortex cells of the pine; some fungi secrete substances which may protect the host from harmful bacteria.
   (c) **Plant cover.** Leaf litter acts as a sponge to hold water and plays an important role in reducing soil erosion and maintaining the water cycle.
   (d) **Timber.** Conifers of the coniferous forests and pine plantations on mountain sides are used as timber in the building industry and pinewood in the production of furniture.

C. Angiosperms

1. Angiosperms, which form the dominant vegetation over the greater part of the earth's surface, are flowering plants since they are the only group of plants which bears **flowers**.
   They are mainly **terrestrial** and well suited to life on land.

2. The adult angiosperm sporophyte plant is differentiated into roots, stems and leaves each with specialised tissues to enable the plant to live successfully on land. These tissues **support** the plant body, **conduct substances** to all parts of the plant and are capable of **producing organic food** by the process of photosynthesis.
   A **complete flower** consists of the calyx, corolla, androecium and gynaecium. **Microspores** develop inside **microsporangia** (pollen sacs) and **megaspores** inside **megasporangia** (nucellus tissue) of the **ovule** as a result of meiosis in **spore mother cells**. A microspore develops into a **pollen grain** consisting of the vegetative cell and generative cell. A megaspore develops into an **embryo sac** (female gametophyte) consisting of an egg cell (ovum), synergids, polar cells and antipodal cells.

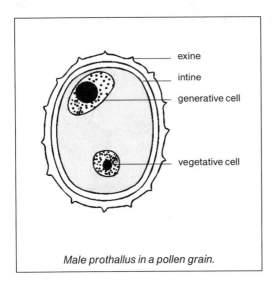

Male prothallus in a pollen grain.

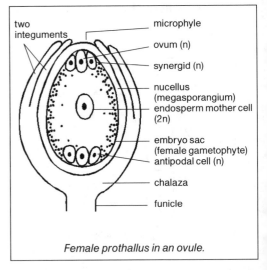

Female prothallus in an ovule.

3. In **pollination** pollen grains are transferred to the stigma from where a **pollen tube** grows from the pollen grain down the style to the ovule. The **two male gametes** are released from the pollen tube into the embryo sac. In angiosperms **double fertilisation** takes place. One male gamete fusing with the egg cell to form a diploid **zygote** and the other with the diploid endosperm mother cell to form a **triploid primary endosperm cell**.
   After fertilisation the zygote develops into an **embryo**, the primary endosperm cell into an **endosperm tissue**, the ovule into a **seed**, the integuments into a **testa** (seed coat), the ovary into a **fruit** and the ovary wall (carpel) into a fruit wall or pericarp. The embryo plant in a seed consists of a **radicle**, one or two **cotyledons** and the **plumule**.

4. The **ecological role** of Angiosperms can be summarised as follows:
   (a) These plants are **autotrophic**, convert radiant energy into chemical potential energy and are **primary producers** in an ecosystem. Since Angiosperms grow, die and decay a constant cycling of inorganic nutrients from the soil to plants and animals and back to the soil is ensured. Without such **cycling of nutrients** an ecosystem would come to a stop and eventually disappear.
   (b) By means of photosynthesis these plants maintain a **favourable oxygen and carbon dioxide balance** in the atmosphere and keep concentrations at constant levels.

5. **Agricultural importance** of Angiosperms can be summarised as follows:
   (a) Parts of plant is used as **food**:
   roots • crops like carrots and sweet potatoes;
   stems • like potatoes and sugar cane;
   leaves • like cabbage and lettuce;
   fruits • like pumpkins, oranges and green beans;
   seeds • like beans and peas;
   (b) Used as **drinks** like coffee and tea.
   (c) Several angiosperms are cultivated for their **medicinal value** like eucalyptus oil, cloves and castor oil.
   (d) Cultivated for **decorative purposes** like flowers and shrubs.
   (e) Serve as **food** for domestic animals e.g. lucerne, hay, lupins, etc.
   (f) Used as **fertilisers**, e.g. leguminous plants to increase soil nitrogen.

STANDARD GRADE
QUESTIONS

Section A

A. *Various possibilities are suggested as answers to the following questions. Indicate the correct answer.*

1. To which of the following groups does a pine-tree belong?
   A   Monocotyledons   C   Conifers
   B   Flowering plants   D   Angiosperms

2. Gymnosperms are
   A   non-vascular seed plants.   C   vascular flowering plants.
   B   seed-bearing vascular plants.   D   seedless vascular plants.

3. Angiosperms and gymnosperms are
   A   seed-producing plants.   C   non-vascular plants.
   B   flowering plants.   D   cone-bearing plants.

4. Which of the following plants produces flowers?
   A   Conifer   C   Grass
   B   Marine alga   D   Pine-tree

5. All cone-bearing plants that produce seeds are classified as
   A   flowering plants.   C   monocotyledons.
   B   angiosperms.   D   gymnosperms.

6. Which of the following plant groups has a heteroptrophic mode of nutrition?
   A   Phycophytes   C   Bryophytes
   B   Mycophytes   D   Pteridophytes

7. Indicate where you would classify a plant lacking chlorophyll.
   A   Spermatophytes   C   Pteridophytes
   B   Phycophytes   D   Mycophytes

8. The female gametophyte in *Pinus* is embedded in the
   A   integuments.   C   nucellus.
   B   ovuliferous scale.   D   megaspore.

9. The pollen grain in *Pinus* is considered to be a
   A   microspore.   C   female gamete.
   B   zygote.   D   male gamete.

10. Which of the following do gymnosperms and angiosperms share?
    A Flowers
    B Seeds
    C Cones
    D Fruit

11. Needle-like leaves of Gymnosperms
    A protect the stem against change of temperature.
    B cannot transpire.
    C do not synthesise organic food.
    D photosynthesise.

12. In Gymnosperms, ova are formed in the
    A antheridia.
    B ovaries.
    C microsporangia.
    D archegonia.

13. Dwarf shoots in *Pinus*
    A have unlimited growth.
    B are replaced by female cones.
    C live for about three years.
    D are leafless.

14. Which of the following cells contains paired homologous chromosomes?
    A Zygote
    B Sperm
    C Gamete
    D Ovum

15. The seeds of a gymnosperm develop on bracts or scales of
    A dwarf shoots.
    B cones.
    C bark
    D flowers.

16. Fertilisation occurs in plants when the pollen grain
    A lands on the stigma.
    B tube grows down the style.
    C tube enters the ovulum.
    D nucleus fuses with the egg cell.

17. Angiosperms are a group of plants which reproduces themselves by the production of
    A spores.
    B seeds.
    C cones.
    D fruit.

18. Pollination is best defined as a process in which
    A pollen unites with an ovule.
    B a male gamete fuses with a female gamete.
    C pollen is received by a receptive stigma.
    D insects carry pollen from one flower to another.

19. All flowering plants are classified as
    A angiosperms.
    B conifers.
    C seed-bearing plants.
    D spermatophytes.

Questions 20 to 23 are based on the following diagram of a flower.

20. Flowers as plant organs are only found in
    A dicots.
    B spermatophytes.
    C monocots.
    D angiosperms.

21. Seeds develop in the part numbered
    A 1.
    B 2.
    C 3.
    D 4.

22. The names of the numbered parts 1 to 5 are respectively
   A   corolla, stigma, stamen, ovary, calyx.
   B   calyx, anther, stamen, pistil, sepal.
   C   petal, stigma, stamen, ovary, corolla.
   D   sepal, pistil, anther, stamen, petal.

23. The reproductive parts of the flower are
   A   1, 2 and 3.
   B   2, 3 and 4.
   C   3, 4 and 5.
   D   2 and 4.

24. The two essential parts of a flower are the
   A   ovary and anther.
   B   corolla and ovary.
   C   pistil and stamens.
   D   calyx and ovary.

25. The four principal structural parts of most flowers are
   A   petals, sepals, pistils, stamens.
   B   calyx, corolla, ovules, pollen.
   C   stamens, pistils, pollen, seeds.
   D   petals, stamens, cotyledons, pistils.

26. What is the relationship between pollination and fertilisation in a flower?
   A   Pollination is sexual and fertilisation asexual
   B   Fertilisation and pollination are the same activity
   C   Pollination must occur before fertilisation can occur
   D   Fertilisation and pollination occur simultaneously

27. The nucellus in Angiosperms represents the
   A   female gametophyte.
   B   megasporangium.
   C   microsporangium.
   D   archegonium.

28. In Angiosperms fertilisation occurs within the
   A   endosperm.
   B   pollen grain.
   C   stigma.
   D   ovule.

29. The number of chromosomes present in the endosperm of a seed of a flowering plant is
   A   haploid
   B   2n.
   C   3n.
   D   diploid.

30. Which of the following does **not** refer to Angiosperms?
   A   The sporophyte is the conspicuous phase
   B   Ovules are enclosed by the ovary
   C   Ovum is located in an archegonium
   D   Reproductive organs are flowers

31. The leaflike part of a plant embryo in a seed is called a
   A   cotyledon.
   B   petiole.
   C   plumule.
   D   sepal.

32. The male gametophyte in Spermatophytes is the
   A   germinated pollen grain.
   B   male gamete.
   C   anther.
   D   nucellus.

33. From which part of a flower does a fruit develop?
   A   Corolla
   B   Anther
   C   Calyx
   D   Pistil

34. Which of the following is found in the embryo sac of flowering plants?
   A   Vegetative nucleus
   B   Synergids
   C   Microspore
   D   Megasporanguim

35. The testa of the seed of Angiosperms develops from the
   A   receptacle.
   B   placenta.
   C   endosperm.
   D   integuments.

Questions 36 to 38 refer to the following diagram.

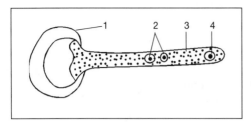

36. The diagram illustrates the
    A development of a hypha in *Rhizopus*.
    B germination of a zygospore in *Spirogyra*.
    C formation of a conjugation tube in Phycophytes.
    D germination of a pollen grain in an angiosperm.

37. The diagram represents a
    A megaspore.   C microspore.
    B male gametophyte.   D germinating zygospore.

38. Which of the following is incorrect?
    A 1 refers to the exine   C 3 refers to the pollen tube
    B 2 refers to the sexual cells   D 4 refers to the generative nucleus

39. After the process of fertilisation in a flower the
    A petals remain permanently attached to the flower.
    B ovules and ovary wall enlarge considerably.
    C whole flower withers and drops off.
    D fruit becomes a seed.

40. Fertilisation is
    A the fusion of the nucleus of a male gamete with the nucleus of a female gamete.
    B when two cells fuse with one another.
    C the transfer of pollen to the stigma.
    D when a sperm enters the egg cell.

41. Which of the following in flowering plants has the most chromosomes per cell?
    A Embryo   C Endosperm
    B Sperm   D Zygote

42. Seed-producing plants are
    A spermatophytes.   C angiosperms and ferns.
    B angiosperms only.   D gymnosperms and dicots only.

43. Which of the following is **not** related to the gynaecium?
    A Ovule   C Stigma
    B Carpel   D Anther

44. Which of the following form(s) the outermost whorl of a flower?
    A Petals   C Calyx
    B Stamens   D Gynaecium

45. Fertilisation in flowering plants occurs in the
    A stigma.   C ovary.
    B ovule.   D pollen tube.

46. A cotyledon is
    A an embryonic shoot.   C an embryonic leaf.
    B the radicle.   D a plant embryo.

B. Write down the correct term for each of the following statements.

1. The hard, resistant outer wall of a pollen grain
2. A branch in *Pinus* with limited growth from which two to three needle-like leaves arise
3. The ripened ovary in a flowering plant
4. The name given to the sterile whorls of a flower
5. A group of plants which produce flowers and ovules within an ovary
6. The tissue representing the megasporangium in *Pinus*
7. The process when one male gamete fuses with an ovum and another gamete with an endosperm mother cell
8. The microsporangia in flowering plants
9. The transfer of ripe pollen from anthers to a receptive stigma
10. The symbiotic association between a fungus and the tissue of a root in a pine
11. The haploid tissue in the ovule of a pine in which the female reproductive organs occur
12. The structure in spermatophytes which, after fertilisation of the ovum inside it, develops into a seed
13. The structure in which fertilisation takes place in a flowering plant
14. A fertilised egg
15. A cavity in an anther in which pollen develops in angiosperms
16. The tough structure into which the integuments develop in a mature seed
17. A modified shoot which serves as sexual reproductive organ in angiosperms
18. The description of a flower having all four whorls of floral parts
19. A ripened ovule after the ovum within it has been fertilised
20. The small opening in the ovule coverings, through which the pollen tube enters during the process of fertilisation.

C. Write down the letter of the description in column B which best suits the term in column A.

Column A		Column B
1. Integument	A	Association of hyphae with roots of pine
2. Gynaecium	B	Branch with limited growth in pine
3. Sporophyte	C	Protects the ovule
4. Mycorrhiza	D	Male reproductive organs of a flower
5. Androecium	E	Female whorl of a flower
6. Dwarf shoot	F	Generation representing the adult gymnosperm plant

D. The following diagram represents a flower of an angiosperm.

Answer the questions.

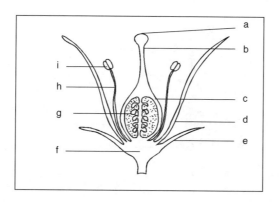

(a) Give **one** reason why the plant of this flower is classified as an Angiosperm.
(b) What is the common name given to parts a, b and c?
(c) Write down the letter of the structure in which the female gamete is produced.
(d) Write down the letters of the parts which represent the androecium.
(e) Which letter refers to the whorl which is important for insect pollination?
(f) Is this flower likely to be self-pollinated or cross-pollinated?
   Give a reason for your answer.
(g) Which letter represents the receptacle?
(h) Which letter represents the part that will develop into:
   (1) a seed and
   (2) a fruit?
(i) Which letter represents the protective whorl while the flower is still in bud?

E. The diagram represents a branch of a pine-tree.

Answer the questions.

(a) Which numbered part

   (i) represents a microstrobilus,
   (ii) bears brown bracts,
   (iii) is a photosynthetic organ, and
   (iv) bears microsporophylls?

(b) Identify the parts numbered 1 to 4.

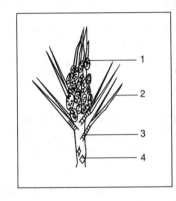

Section B

1. (a) What is meant by the term Spermatophytes?
   (b) Name the (i) division, (ii) sub-division and (iii) genus to which the pine belongs.
   (c) In which ways are the pine-tree adapted for wind pollination?
   (d) Briefly describe the habitat of the pine tree.

2. Describe in what ways each of the following parts of a pine are well suited to a terrestrial mode of life:
   (a) **roots;**
   (b) **stem;**
   (c) **leaves;** and
   (d) **seed.**

3. The diagram shows part of a branch of *Pinus* sp.

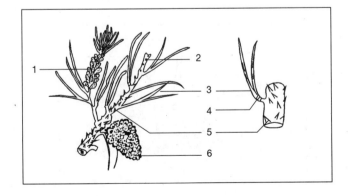

(a) Identify the parts numbered 4 and 5.
(b) Write down the number of the part in which:
   (i) microspores are produced;
   (ii) seed is produced;
   (iii) photosynthesis takes place; and
   (iv) unlimited growth can take place.
(c) What is the average life-span of the part numbered 4?
(d) How does the male gamete from the pollen grain reach the ovum?
(e) How long after pollination will fertilisation take place in *Pinus* sp.?
(f) State the most important difference, as far as position is concerned, between the ovule of *Pinus* sp. and the ovule of a flowering plant.
(g) Draw a labelled diagram to illustrate the position and internal structure of the mature ovule in *Pinus* sp.

4. Study the diagram of a longitudinal section through an ovule of *Pinus*.

   (a) Identify the parts numbered 1 to 9.
   (b) Which number indicates the:
      (i) megasporangium,
      (ii) female gametophyte, and
      (iii) male gametophyte?
   (c) Write down the numbers of those parts which are haploid.
   (d) State *three* reasons as illustrated in the diagram, why it is and ovule of a gymnosperm.
   (e) How long after pollination does fertilisation take place in the pine?
   (f) How are the male gametes from the male gametophyte transported to the ovum?

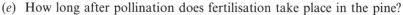

5. Answer the following questions regarding Spermatophytes.
   (a) Distinguish between the concepts pollination and fertilisation in flowering plants.
   (b) Name the main floral parts of a complete flower.
   (c) Name the reproductive organs of
      (i) Angiosperms, and
      (ii) Gymnosperms.

6. Briefly describe the ecological role of Gymnosperms under the following headings:
   (a) **effect on soil**, (b) **mycorrhiza**, (c) **plant cover**, (d) **timber**.

7. This diagram represents a section through the anther of a flower.

   (a) Identify the structure labelled 4.
   (b) Identify the structure labelled 3.
       By what process is this structure produced?
   (c) Describe how the structure labelled 3 receives nourishment. Use **numbers** and **names** in your answer.

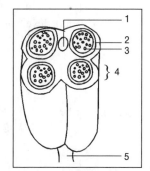

8. The diagram represents a longitudinal section through the gynaecium of a flowering plant prior to fertilisation.

   Study it and answer the questions.

   (a) Identify the parts numbered 1 to 16.

59

(b) Write down the numbers of **five** haploid cells.
(c) Write down the number of **one** diploid cell.
(d) Write down the number which represents the
   (i) male gametophyte,
   (ii) female gametophyte, and
   (iii) megasporangium.
(e) Briefly describe the double fertilisation which will occur in this gynaecium.
(f) What is the advantage of double fertilisation to a plant?
(g) What happens to the cells numbered 14 and 16 after fertilisation?
(h) Which part of the seed develops from the structure numbered
   (i) 8,
   (ii) 13, and
   (iii) 15?

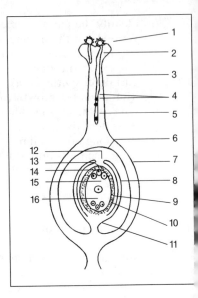

9. Answer the questions on the diagram of a germinating pollen grain in Angiosperms.

   (a) Identify the parts numbered 1 to 4.
   (b) State the functions of the parts numbered:
      (i) 3, and
      (ii) 4.

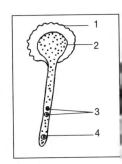

10. The following diagram represents an ovule of a flower.
   (a) Write down the number which represents the:
      1. ovum;
      2. cell which will become triploid after fertilisation;
      3. megasporangium;
      4. female gametophyte; and
      5. antipodal cell.
   (b) Identify the following parts:
      4;   7;   8
   (c) What develops from each of the following parts?
      3;   5;   6

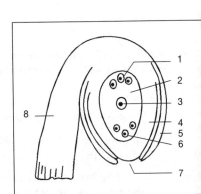

11. Describe the (a) ecological role, and
                 (b) agricultural importance of Angiosperms.

STANDARD GRADE
ANSWERS

Section A

A. 1. C 2. B 3. A 4. C 5. D 6. B 7. D
    8. C 9. A 10. B 11. D 12. D 13. C 14. A
    15. B 16. D 17. B 18. C 19. A 20. D 21. D
    22. A 23. B 24. C 25. A 26. C 27. B 28. D
    29. C 30. C 31. A 32. A 33. D 34. B 35. D
    36. D 37. B 38. D 39. B 40. A 41. C 42. A
    43. D 44. C 45. B 46. C

B. 1. Exine 2. Dwarf shoot 3. Fruit
    4. Perianth 5. Angiosperm 6. Nucellus
    7. Double fertilisation 8. Pollen sacs 9. Pollination
    10. Mycorrhiza 11. Embryo sac/female gametophyte 12. Ovule
    13. Embryo sac 14. Zygote 15. Pollen sac
    16. Testa 17. Flower 18. Complete
    19. Seed 20. Micropyle

C. 1. C 2. E 3. F 4. A 5. D 6. B

D. (a) Ovules are enclosed in an ovary
   (b) Pistil (c) g (d) h and i (e) d
   (f) Cross-pollination; stigma grows higher than anthers
   (g) f (h) (1) g (2) c (i) e

E. (a) (i) No. 1 (ii) No. 3 (iii) No. 2 (iv) No. 1
   (b) 1 – male cone 2 – leaf 3 – dwarf shoot 4 – scar of dwarf shoot

Section B

1. (a) Plants bearing seeds
   (b) (i) Spermatophytes (ii) Gymnosperms (iii) *Pinus*
   (c) • Large quantities of pollen produced
       • Pollen grains are smooth and light
       • Pollen grain has two air sacs for floating in wind
       • Ovules in female cone are exposed and naked
       • Male cones are produced high on tree near tip of shoots
   (d) Is terrestrial; infertile sandy soil; dry, cold and windy areas

2. (a) **Roots**
       Lateral roots branch close to soil surface
       Allows rapid uptake of surface water present for a short time
       Root hairs are absent and function is taken over by mycorrhiza
       Hyphae of fungal mycelium penetrate between cortex cells of root
       Water and salts from soil are transported by hyphae to cortex

(b) **Stem**
Stem is protected by bark (rhytidome)
Trunk is tall to bear lateral branches over a large area
for leaves to absorb maximum amount sunlight for photosynthesis
Contain strengthening tissue to withstand effect of wind and
vertical mass of the above-ground parts like leaves and branches
Contains conducting tissue for transport of water and salts to leaves
Wounds in damaged stems are sealed by resin

(c) **Leaves**
Needles are thin and reduces loss of water and
risk of damage from cold, ice and snow
Stomata are deeply sunken in the epidermis
which helps to reduce water loss
Epidermis is covered by a cuticle
Leaf surface is much reduced to limit water loss

(d) **Seed**
Is surrounded by tough, waterproof testa
which protects seed from drying out in dormant stage
Has a supply of stored food for its initial growth
Has a papery wing for dispersal by wind

3. (a) 4 – dwarf shoot   5 – scale leaf
   (b) (i) 1   (ii) 6   (iii) 3   (iv) 2   (c) 3 years
   (d) Moves along pollen tube; to archegonium in female gametophyte
   (e) One year
   (f) In *Pinus* it is borne naked on ovuliferous scale of cone;
       In flowering plant it is enclosed in an ovary of flower

(g)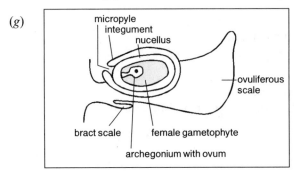

4. (a) 1 – micropyle   2 – integument   3 – nucellus/megasporangium
       4 – female prothallus 5 – ovuliferous scale 6 – archegonium
       7 – ovum   8 – pollen tube   9 – bract scale
   (b) (i) 3 (ii) 4 (iii) 8
   (c) 4, 6, 7, 8
   (d) Presence of archegonium in female prothallus
       Pollen tube grows directly into nucellus
       Ovule is borned naked
   (e) One year
   (f) Along pollen tube

5. (a) **Pollination** – transfer of ripe pollen grains from an anther to the receptive stigma of the same flower or another of the same species
       Fertilisation – fusion of a male gamete with an ovum to produce a fertilised egg or zygote
   (b) Calyx; Corolla; Stamens; Pistil
   (c) (i) Flowers   (ii) Cones

6. (a) **Effect on soil**
   - Fallen pine leaves decompose slowly
   - Humic acids accumulate and few humus is formed
   - Soil becomes infertile
   (b) **Mycorrhiza**
   - Symbiotic relationship exists between pine roots and fungal mycelium
   - Water and minerals are transported by hyphae to cortex cells
   - Fungi secrete substances to protect host from harmful bacteria
   (c) **Plant cover**
   - Leaf litter acts as sponge to hold water
   - Plays role in reducing soil erosion
   - And to maintain water cycle
   (d) **Timber**
   - Are used as timber in building industry
   - And in production of furniture

7. (a) Anther lobe   (b) Microspore/pollen grain; meiosis
   (c) Nutrients are transported along thin filament No. 5; to anther; from vascular bundle No. 1; nutrients diffuse to pollen sac No. 2

8. (a)
   1 – stigma   2 – pollen tube   3 – style
   4 – male gametes   5 – tube nucleus   6 – loculus
   7 – ovary wall   8 – integuments   9 – embryosac
   10 – nucellus (megasporanguim)   11 – funicle   12 – micropyle
   13 – ovum   14 – synergid   15 – endosperm mother cell
   16 – antipodal cells
   (b) 4, 5, 13, 14, 16 (c) 15
   (d) (i) 2 (ii) 9 (iii) 10
   (e) **double fertilisation**:
   Tip of pollen tube opens in embryo sac
   Two male gametes are released
   One male gamete fuses with nucleus of ovum
   To form a diploid zygote
   Other male gamete fuses with two polar nuclei/endosperm mother cell
   To form triploid primary endosperm cell
   (f) **Advantage**: Endosperm serves to provide nutrition
                      To young embryonic plant
                          During its critical early stages of development in a terrestrial environment
   (g) Cells disappear
   (h) (i) Integuments form the testa of the seed
       (ii) Zygote develops into embryo
       (iii) Triploid nucleus develops into endosperm

9. (a) 1 – exine   2 – intine   3 – male gametes   4 – vegetative nucleus
   (b) (i) One gamete fuses with ovum in embryo sac; to form diploid zygote
           Other gamete fuses with diploid endosperm mother cell; to form triploid primary endosperm cell
       (ii) Controls pollen tube's direction of growth

10. (a) 1 – No. 6   2 – No. 3   3 – No. 2   4 – No. 2   5 – No. 1
    (b) 4 – nucellus   7 – micropyle   8 – funicle
    (c) 3 – endosperm   5 – testa   6 – embryo

11. (a) Angiosperms are autotrophic and produce own organic food
        By photosynthesis they produce own organic food for themselves

As producers they serve as food for other organisms
They serve as first trophic level in food chains
They support primary consumers and indirectly secondary consumers

(b)
- Parts of plant is used as food for man, e.g. carrots, potatoes, etc.
- Used as drinks like coffee and tea
- Cultivated for their medicinal value like eucalyptus oil
- Cultivated for decorative purposes like flowers and shrubs
- Serve as food for domestic animals, e.g. lucerne and hay
- Used as fertilisers, e.g. legumes

HIGHER GRADE
QUESTIONS

Section A

A. *Various possibilities are suggested as answers to the following questions. Indicate the correct answer.*

1. In the life cycle of *Pinus* the nucellus is the
   A   megaspore.
   B   microspore.
   C   female gametophyte.
   D   megasporangium.

2. The chromosome number of the nuclei in the endosperm of gymnosperms is
   A   diploid.
   B   n.
   C   2n.
   D   3n.

3. Which of the following forms part of the gametophyte generation in gymnosperms?
   A   Nucellus
   B   Megasporophyll
   C   Endosperm
   D   Megaspore mother cells

4. Which of the following must occur for a sporophyte to produce spores?
   A   Meiosis
   B   Pollination
   C   Germination
   D   Spore dispersal

5. An ovule is to a seed, as an ovary is to a(an)
   A   ovule.
   B   pollen grain.
   C   sporophyll.
   D   fruit.

6. Which of the following describe sexual reproduction best? The production of offspring
   A   from one parent.
   B   from two parents.
   C   by the fusion of two zygotes.
   D   by the fusion of two gametes.

7. In gymnosperms the sperms are carried to the ovum by
   A   ovules.
   B   pollen tubes.
   C   nucellus.
   D   water.

8. Which of the following is **not** true with regard to the branches of *Pinus* sp.?
   A   Lateral branches are carried as a series of whorls around stems
   B   Dwarf shoots develop in the axils of bracts
   C   Lateral branches have limited growth
   D   Dwarf shoots also grow on the lateral branches

9. The correct sequence of five phases in the life cycle of *Pinus* sp. is
   A   microspore, zygote, male gamete, embryo, tree.
   B   microspore, embryo, tree, male gamete, zygote.
   C   microspore, male gamete, zygote, embryo, tree.
   D   microspore, embryo, zygote, male gamete, tree.

10. In Gymnosperms the
    A   ovule is borne naked on an ovuliferous scale.
    B   pollen grains fall on the stigma.
    C   pollen tube grows down the style to the ovary.
    D   endosperm is formed after fertilisation.

11. Which of the following belongs to the sporophyte generation?
    A   Microspores in Spermatophytes   C   Megaspores in Angiosperms
    B   Archegonium in Gymnosperms   D   Sporogonium in Bryophytes

12. Which of the following is the function of the vegetative nucleus of the pollen grain?
    A   Fuses with the ovum   C   Fuses with the endosperm mother cell
    B   Gives rise to an embryo   D   Controls the direction of growth of the pollen tube

13. A seed and a spore differ in that
    A   spores are diploid and seeds are haploid.
    B   the spores only can be unicellular.
    C   spores can withstand dehydration while seeds cannot.
    D   spores are gametes whilst seeds give rise to new plants.

14. The microspore mother cells giving rise to pollen grains in *Pinus* sp. and in Angiosperms, are similar in that they both
    A   are haploid.   C   undergo meiosis.
    B   developed in anthers.   D   are microspores.

15. Meiosis in *Pinus* sp. occurs
    A   during the production of spores.   C   prior to fertilisation.
    B   during the production of gametes. D   after the production of spores.

16. The megasporangium in *Pinus* is the
    A   embryo sac.   C   archegonium.
    B   ovule.   D   nucellus.

Questions 17 to 20 refer to the following diagram of the life cycle of an angiosperm.

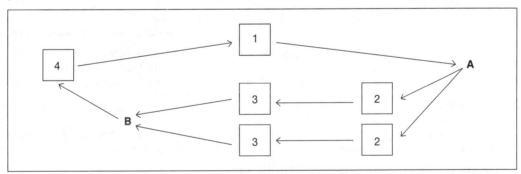

17. If number 1 represents the adult angiosperm and 2 its microscopic gametophyte, then A must be the
    A   spores.   C   seed.
    B   gametes.   D   zygote.

18. If the two number 2's are microscopic gametophytes in the flower then the two 3's are
    A   seeds and zygotes.   C   spores and pollen.
    B   ova and male gametes.   D   two embryos in the seed.

19. B restores the diploid number of chromosomes in the
    A   zygote.   C   seed.
    B   gametes.   D   spores.

65

20. The structure numbered 4, which develops into the adult angiosperm at 1, is the
   A   zygote.   C   seed.
   B   spore.   D   endosperm.

21. Which *three* of the following make up the embryo of a seed?
   1. **endosperm**   2. **plumule**   3. **pericarp**   4. **cotyledon**   5. **testa**   6. **radicle**
   A   3, 4, 5   C   1, 2, 4
   B   1, 2 6   D   2, 4, 6

22. In which of the following parts of a flowering plant does meiosis occur?
   A   Cotyledons   C   Pollen grains
   B   Anthers   D   Pollen tube

23. Each of the following is a characteristic of Angiosperms except that they all
   A   bear ovules in an ovary.
   B   have a separate calyx and corolla.
   C   have either one or two cotyledons.
   D   have a prominent diploid stage in their life cycle.

24. The endosperm of the ovule in Angiosperms is the
   A   product of pollination.   C   diploid tissue of the ovary.
   B   diploid product of fertilisation.   D   nutritive tissue of the embryo.

25. Double fertilisation in Angiosperms results in
   A   two cotyledons developing in the embryo.
   B   a triploid zygote and diploid endosperm cell.
   C   a diploid zygote and a triploid endosperm cell.
   D   fusion of the ovum and a primary endosperm nucleus.

26. Which of the following is a process of sexual reproduction in plants?
   A   Spore formation in fungi   C   Seed formation in the pea
   B   Spore formation in ferns   D   Spore formation in mosses

27. In many cases plants can be more pure in their reproduction than animals. The reason for this is because they
   A   reproduce by means of seeds.   C   are cross-pollinated.
   B   reproduce vegetatively.   D   are more stable as far as their genes are concerned.

28. Angiosperms are different from Gymnosperms in that in Gymnosperms
   A   water is not required for fertilisation.   C   double fertilisation occurs.
   B   the endosperm is formed before fertilisation.   D   the sporophyte generation is dominant.

29. If the nuclei of the pollen grains of a gymnosperm contain 20 chromosomes, there will be 20 chromosomes in the nuclei of
   A   egg cells.   C   ovuliferous scales of the female cones.
   B   cells of the cortex of the stem.   D   cells of nucellus tissue.

Questions 30 to 32 refer to the following diagram.

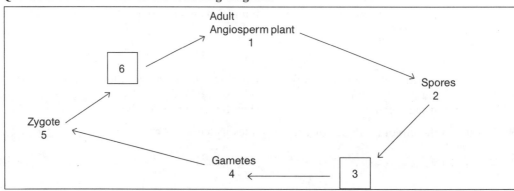

30. The dominant phase in the life cycle of the angiosperm plant is represented by number
 A 1.   C 5.
 B 3.   D 6.

31. Pollen grains, which are produced, are represented by number
 A 2.   C 4.
 B 3.   D 6.

32. Meiosis takes place between numbers
 A 1 and 2.   C 3 and 4.
 B 2 and 3.   D 5 and 6.

33. The female gametophyte is represented by number
 A 1.   C 3.
 B 2.   D 6.

Questions 34 and 35 refer to the following diagram.

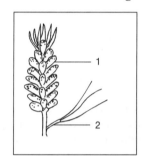

34. Structure 1
 A represents a female cone.
 B drops off after three years and germinates.
 C is a long branch of unlimited growth.
 D develops in the place of a dwarf shoot.

35. Structure 2
 A gives rise to a female cone.
 B is a dwarf shoot with limited growth.
 C develops in the place of a lateral branch.
 D is the origin of a long branch with unlimited growth.

Questions 36 to 38 refer to the schematic representation of the life cycle of *Pinus*.

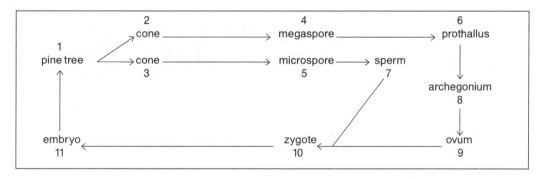

36. Meiosis occurs between numbers
 A 1 and 2.   C 8 and 9.
 B 2 and 4.   D 9 and 10.

37. The diploid phase is represented by
 A 1, 2, 3, 6 and 10.   C 2, 3, 4, 6 and 8.
 B 4, 6, 7, 8 and 9.   D 1, 2, 3, 10 and 11.

38. In this life cycle the only parts which undergo/undergoes dispersal from place to another, is/are
   A 2 and 3.
   B 3 and 5.
   C 5 and 11.
   D 4 and 6.

39. How does the gametophyte generation of a plant species differ from the sporophyte generation?
   A It produces the spores
   B It is diploid in all its cells
   C It is always dependent on the sporophyte
   D It is haploid in all its cells

40. The seeds and spores of Spermatophytes differ in that
   A spores are produced sexually and seeds asexually.
   B seeds are embryos and spores zygotes.
   C spores are much larger than seeds.
   D seeds are diploid and spores haploid.

B. Write down the correct term for each of the following statements.

1. The female gametophyte in spermatophytes
2. A coat surrounding an ovule in angiosperms
3. The symbiotic relationship of fungal hyphae, e.g. *Amanita muscaria* with the roots of *Pinus*
4. A diploid cell in a sporangium that gives rise to spores by meiosis
5. The whorl of stamens in flowering plants
6. The triploid nutritive tissue outside the embryo in the seed
7. The pistil in flowering plants
8. The production of two different kind of spores known as microspores and megaspores
9. The short stalk found in angiosperms which attaches an ovule to the placenta
10. The small opening in the ovule coverings through which the pollen tube usually enters during the process of fertilisation
11. The megasporangium tissue of an ovule which encloses the female gametophyte in spermatophytes
12. The calyx and the corolla of a flower taken together
13. The ripened ovary wall of a fruit
14. Embryonic or first leaf of angiosperms and gymnosperms and which differs from the subsequent leaf
15. Cells which fuse to form a zygote
16. A flower which has both an androecium and a gynaecium
17. The name given to leaf producing male spores
18. The generation which is represented by the adult gymnosperm plant

C. The diagram below shows the structure of a mature embryo sac in an ovule of a flowering plant. This plant has 20 chromosomes in the nucleus of each cell making up the leaves of the plant. Opposite each of the numbers in the diagram, write down the name of the part and the relevant chromosome number of the cell(s) making up the part.

	Name of part	Chromosome number
1.		
2.		
3.		
4.		
5.		
6.		
7.		

D. **Select from column B the plant group which can best be associated with the characteristic in column A.**

Column A
1. The ovule is enclosed in an ovary
2. The mature plant is the gametophyte
3. A columella is formed during asexual reproduction
4. Ovule contains archegonia
5. Haploid spore gives rise to a prothallus
6. Ribbon-like chloroplasts are present
7. Hyphae secrete diastase which hydrolyses starch into soluble sugars
8. The sporophyte is known as a sporogonium
9. The gametophyte is a thallus and independent on the sporophyte
10. Thallus plants with no chlorophyll

Colom B

A Pteridophytes
B Gymnosperme
C Bryophytes
D Mycophytes
E Angiosperms
F Phycophytes

Section B

1. Answer the following questions regarding gymnosperms.
   (a) Which cell in the young ovule gives rise to the female gametophyte?
   (b) What is the chromosome composition of the cell mentioned in (a)?
   (c) By which process does the cell mentioned in (a) give rise to the female gametophyte?
   (d) When does fertilisation occur in *Pinus* sp?
   (e) Briefly describe the fertilisation process in *Pinus* sp.

2. Answer the following questions regarding the pine.
   (a) Where are the male cones found on a pine tree?
   (b) Draw a labelled diagram of a longitudinal section of the mature male cone of the pine to illustrate its structure.
   (c) What is the name given to the male spores in the pine?
   Describe where and how these spores are formed.
   (d) In what ways are the archegonia of the pine different from that of mosses?

3. The diagrams below show two types of branches found in a group of plants.

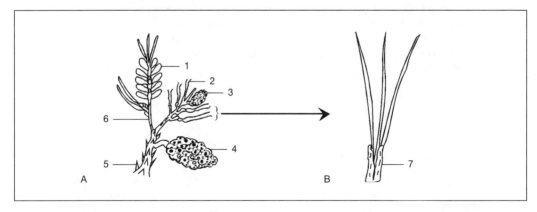

   (a) Name the: (i) division, and   (ii) sub-division to which this group of plants belong.
   (b) State **three** visible reasons for your answer in (a) (ii).
   (c) Identify shoots A and B respectively.
   (d) Identify the parts numbered 1 to 7.

4. The diagrams below represent the life cycle of *Pinus* sp. Study these diagrams and answer the questions.

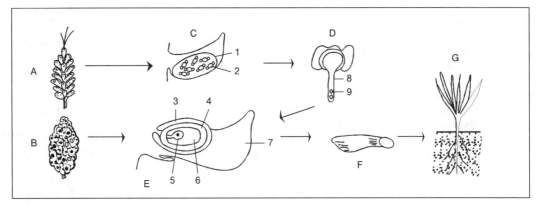

(a) If an ovum of *Pinus* sp. has **20** chromosomes in its nucleus, how many chromosomes are there in the nuclei of the cells found in the following structures?
   (i) F (ii) G (iii) 4 (iv) 6 (v) 9
(b) Which of the structures numbered A, B, D F or G represents the young sporophyte?
(c) Which of the numbered parts presents the:
   (i) microsporangium;
   (ii) megasporangium;
   (iii) female gametophyte;
   (iv) male gametophyte;
   (v) ripened ovule; and
   (vi) megasporophyll?
(d) Is *Pinus* sp. monoecious or dioecious? Give a reason.

5. Compare in tabular form the life cycle of Pteridophytes and Angiosperms with regard to:
   (a) sporangia;
   (b) types of spores;
   (c) gametophytes; and
   (d) transfer of gametes.

6. Complete the following diagram of the life cycle of a flowering plant by writing down the letters of the following terms in the appropriate blocks numbered 1 to 7.
   A pollen tube growth   E seed germination
   B seed formation   F pollination
   C plant growth   G seed dispersal
   D fertilisation

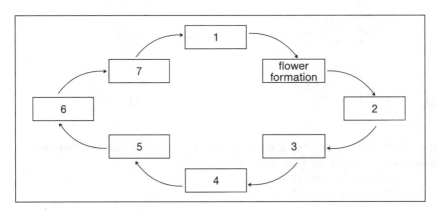

7. (a) Explain the meaning of the terms haploid and diploid when they are used in connection with the nucleus of a cell.
   (b) In which generation of the life cycle, and where precisely does meiosis take place in each of the following?
      1. Angiosperms   2. Gymnosperms   3. Pteridophytes   4. Bryophytes
   (c) Where in the body of a mammal, and when does meiosis take place?
8. Describe what happens from the moment that a ripe pollen grain lands on the susceptible stigma of a flower untill fertilisation has been completed.
9. The diagrams below represent the life cycle of an angiosperm. Study these diagrams and answer the questions.

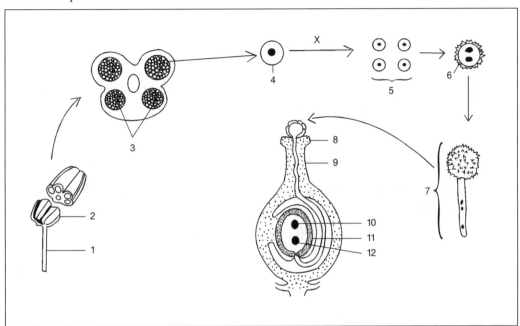

   (a) Identify the parts numbered 1 to 12.
   (b) This angiosperm has 18 chromosomes in the nucleus of each cell making up the leaves of the plant. What will be the chromosome number in the part labelled
      (i) 4   (ii) 6   (iii) 8   (iv) 10   (v) 12?
   (c) What process is taking place at X?
   (d) Which numbered part represents the   (i) megasporangium,
      (ii) microsporangium, and   (iii) male gametophyte?
10. Compare in tabular form the gametophyte generation of a moss with the gametophyte generation of a flowering plant.
11. The following diagram represents a section through a portion of an angiosperm gynaecium. Answer the questions using relevant numbers and names.

    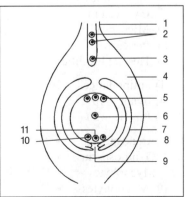

    (a) Identify the parts numbered 2 and 3.
       Explain how these parts were formed in a pollen grain.
    (b) Write down the names of the parts making up the female gametophyte. Opposite each name indicate whether the part is haploid or diploid
    (c) How does the female gametophyte obtain the necessary nutrients for its development?
    (d) From observation of the diagram, what features indicate that it is a structure of an angiosperm and not of a gymnosperm

12. State a difference between:
    (a) the position of the ovule in angiosperms and that in gymnosperms
    (b) the position of the ovum in a flowering plant and that in a pine;
    (c) the number of cotyledons of angiosperms and that of gymnosperms; and
    (d) the chromosome composition of the endosperm in a flowering plant and that in a pine.

13. Tabulate **four** differences between Angiosperms and Gymnosperms regarding
    (a) the female gametophytes and cotyledons, and
    (b) sexual reproductive processes.

Section C

1. Compare the alternation of generations of a fern and an angiosperm with regard to the following:
    (a) where the sporangia are found;
    (b) the production of spores and the further development of the spores into a gametophyte;
    (c) the nutrition of the gametophyte; and
    (d) how fertilisation is brought about.

2. Answer the following questions on *Pinus*:
    (a) Where does the megaspore mother cell develop? Describe the development of this cell that leads to the production of the female gamete.
    (b) Describe the three different processes that have to take place before this ovum develops into an adult seed to be dispersed by wind.

HIGHER GRADE
Answers

Section A

A. 1. D   2. B   3. C   4. A   5. D   6. D   7. B
    8. C   9. C   10. A   11. D   12. D   13. B   14. C
    15. A   16. D   17. A   18. B   19. A   20. C   21. D
    22. B   23. B   24. D   25. C   26. C   27. B   28. B
    29. A   30. A   31. B   32. A   33. C   34. D   35. B
    36. B   37. D   38. C   39. D   40. D

B. 1. Embryo sac   2. Integument   3. Mycorrhiza
    4. Spore mother cell   5. Androecium   6. Endosperm
    7. Gynaecium   8. Heterosporous   9. Funicle
    10. Micropyle   11. Nucellus   12. Perianth
    13. Pericarp   14. Cotyledon   15. Gametes
    16. Hermaphroditic   17. Microsporophyll   18. Sporophyte

C. (1) antipodal cell – 10   (2) nucellus – 20   (3) polar cell – 10
    (4) ovum – 10   (5) synergid – 10   (6) integument – 20
    (7) funicle – 20

D. 1. Angiosperms • E   2. Bryophytes • C   3. Mycophytes • D
    4. Gymnosperms • B   5. Pteridophytes • A   6. Phycophytes • F
    7. Mycophytes • D   8. Bryophytes • C   9. Pteridophytes • A/C
    10. Mycophytes • D

Section B.

1. (a) Megaspore mother cell   (b) Diploid   (c) Meiosis
   (d) One year after pollination
   (e) Two male gametes move along pollen tube
       Towards female gamete (ovum) in archegonium of female gametophyte
       Tube grows down the neck canal of archegonium
       Tip opens and one gamete fuses with ovum
       To form diploid zygote and other gamete disintegrates

2. (a) Immediately behind the terminal bud of lateral shoots in clusters; where they replace dwarf shoots
   (b)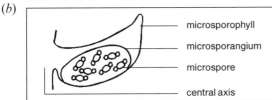
   (c) Microspores
   - on underside of microsporophyll
   - are two microsporangia
   - in which are diploid microspore mother cells
   - each divides meiotically
   - to give rise to four haploid single celled microspores
   - at a later stage each develops an exine and two air sacs

   (d) **Pine**
   1. 2 to 3 archegonia
   2. Archegonia are small
   3. Neck is very short
   4. Venter sunken into prothallus
   5. Stalk is absent

   Moss
   1. A number of archegonia
   2. Archegonia are larger
   3. Long neck
   4. At tip of gametophyte
   5. Long stalk is present

3. (a) (i) Spermatophytes (ii) Gymnosperms
   (b) Reproduction organs are cones
       Long shoots and dwarf shoots are present
       Leaves are needle-like
   (c) A – long shoot   B – dwarf shoot
   (d) 1 – microstrobilus (male cone)   2 – needle-like leaf
       3 – megastrobilus (female cone)   4 – female cone
       5 – bract (scale leaf)   6 – branch with unlimited growth
       7 – bract

4. (a) (i) 40 (ii) 40 (iii) 40 (iv) 20 (v) 20
   (b) G
   (c) (i) 1 (ii) 4 (iii) 6 (iv) D
       (v) F (vi) 7
   (d) Monoecious • male and female sex organs / cones are produced on the same tree

5. **Pteridophytes**
   (a) One type of sporangium in sorus
       On abaxial side of frond

   (b) One type of spore
       Therefore homosporous

   Angiosperms
   Two types of sporangia
   (i) Microsporangia i.e. pollen sacs in anthers of stamens
   (ii) Megasporangia i.e. nucellus tissue in the ovule

   Two types of spores
   Therefore heterosporous
   (i) microspore and   (i) megaspore

(c) **Pteridophytes**
Composed of many cells
Heart-shaped green structure known as prothallus
With archegonia and antheridia
Is independent plant
Anchored by rhizoids

Angiosperms
Very much reduced
Male gametophyte i.e. germinating pollen grain has two cells only
Female gametophyte i.e. embryo sac in ovule composed of eight cells
No antheridia or archegonia
Dependent on sporophyte

(d) Water is required
Sperm swim towards ovum

No water is required
Gametes move along pollen tube

6. 1. C 2. F 3. A 4. D 5. B 6. G 7. E

7. (a) **Haploid:** the n number of chromosomes; characteristic of gametes; half the number of chromosomes in nucleus of somatic cell
   Diploid: 2n number of chromosomes; this is double set of chromosomes; twice that of a gamete of a given species
   (b) 1. **Angiosperms:** sporophyte • in pollen sacs of anthers; and in nucellus tissue of ovule in the flower
   2. **Gymnosperms:** sporophyte • in pollen sacs of male cones; and in nucellus tissue of ovule in female cone
   3. **Pteridophytes:** sporophyte • in sporangium of sorus on frond
   4. **Bryophytes:** sporophyte • in sporangium of sporogonium
   (c) Gonads – testes and ovaries; during gametogenesis

8. Exine of pollen grain softens and intine grows out to form pollen tube
Tube nucleus is at tip of tube and controls direction of growth
Generative nucleus divides mitotically to form two male gametes
Pollen tube grows into style
When it reaches ovule, the pollen tube grows through micropyle
Into nucellus and embryo sac
Tube nucleus disintegrates
Tip of pollen tube ruptures and two male gametes are released into embryo sac
One of them fuses with ovum to form a diploid zygote
The other one fuses with diploid endosperm mother cell
To form a triploid primary endosperm cell
Double fertilisation takes place

9. (a) 1 – filament   2 – anther   3 – microsporangium
       4 – microspore mother cell   5 – microspores   6 – pollen grain
       7 – germinating pollen grain 8 – stigma   9 – style
       10 – primary endosperm cell   11 – nucellus   12 – zygote
   (b) (i) 18 (ii) 9 (iii) 18 (iv) 27 (v) 18
   (c) Meiosis
   (d) (i) 11 (ii) 3 (iii) 7

10. **Moss**
1. adult plant is the gametophyte
2. clearly observable
3. conspicuous prominent generation
4. gametophyte is an independent plant
5. possesses antheridia and archegonia
6. reproductive structures receive less protection

Angiosperm
– germinating pollen grain is the male and embryo sac female gametophyte
– not observable inside ovule
– are reduced to a few nuclei only
– entirely dependent on sporophyte for protection and nutrition
– possesses no antheridia and archegonia
– high degree of protection is provided to essential reproductive structures

11. (*a*) 2 – male gametes 3 – tube nucleus
Pollen grain is the male gametophyte
Haploid nucleus of microspore divides by mitosis to give rise to two cells
One is the vegetatives cell known as the tube nucleus
And the other the generative cell
That divides mitotically to form two sperm cells/male gametes
(*b*) 5 – antipodal cells – n
6 – endosperm mother cell – 2n
10 – synergid – n
11 – ovum – n
(*c*) • dependent on sporophyte parent
• uses nutrients from nucellus tissue
• obtained from vascular bundles via funicle
(*d*) Style No. 1 is present
Ovule is enclosed in ovary No. 4
Ovum No. 11 is not in an archegonium
Tube nucleus No. 3 is present
Double fertilisation takes place

12. (*a*) **ovule:** angiosperms – is enclosed in an ovary of flower
gymnosperms – is borne naked on megasporophyll of cone
(*b*) **ovum:** flower – inside embryo sac or female gametophyte
pine – inside archegonium of female gametophyte
(*c*) **cotyledons:** angiosperms – one or two
gymnosperms – five to eight
(*d*) **chromosome number:** flowering plant – triploid (3n)
pine – haploid (n)

13. Angiosperms   **Gimnosperms**

(*a*) Ovule is enclosed within the ovary of flower   Ovule developes naked on ovuliferous scale of cone
Ovum situated inside embryo sac   Ovum situated inside archegonium of female gametophyte

Endosperm is triploid   Endosperm is haploid
One or two cotyledons   Five to eight cotyledons

(*b*) Are pollinated by insects, water and wind   Are pollinated by wind
Pollen grain lands onto stigma   Pollen grain lands directly onto nucellus of ovule

Double fertilisation occurs   Single fertilisation occurs
Endosperm is a product of fertilisation   Endosperm produced prior to fertilisation

Section C

1. (*a*) **Where sporangia are found**
Fern: • sori are situated
• on abaxial side of pinnule of frond
• attached to thickened part known as the placenta
• from where stalked sporangia arise
• which are completely covered by indusium
angiosperm: • microsporangia are the pollen sacs
• in the anthers of the flower
• megasporangia are the nucellus
• in the ovules of flowers

(b) **Development of gametophyte**
  fern:
  - sporogenous tissue of sporangium
  - divides mitotically into diploid spore mother cells
  - each undergoes meiosis to form tetrad of haploid spores
  - which are liberated by action of annulus
  - and germinate in cool, moist, sheltered places
  - to form the gametophyte known as prothallus
  - with antheridia and archegonia on ventral side

  angiosperm:
  - microspore mother cells in pollen sacs
  - divide meiotically to form haploid microspores
  - which develop into pollen grains (male gametophyte)
  - after pollination the exine softens and
  - intine grows out to form pollen tube
  - containing vegetative nucleus and small generative nucleus
  - the latter divides into two male gametes
  - germinating pollen tube with content is mature male gametophyte
  - pollen tube grows down into style through micropyle of ovule
  - into nucellus and embryo sac
  - in ovule one nucellus cell divides by meiosis
  - to form four haploid megaspores
  - the three magaspores nearest micropyle disintegrate
  - centrally placed megaspore gives rise to
  - embryo sac, which is the female gametophyte
  - consisting of ovum, synergids and antipodal cells
  - polar nuclei fuse to form endosperm mother cell

(c) **Nutrition of gametophyte**
  fern:
  - prothallus contains chlorophyll
  - and photosynthesises own carbohydrates
  - is an independent plant
  - rhizoids absorb mineral ions and water

  angiosperm:
  - gametophyte completely dependent on sporophyte
  - male gametophyte obtains nutrients by tapetum cells
  - which received them via vascular bundles
  - female gametophyte in ovule uses nutrients
  - from nucellus tissue which obtains them via the vascular bundles in funicle

(d) **Fertilisation**
  fern:
  - neck canal cells of archegonium
  - secrete slimy substance containing malic acid
  - antheridia burst at top and male gametes are liberated
  - which are attracted by malic acid in slimy secretion
  - swim in film of water along neck canal
  - fuses with ovum in venter to form diploid zygote

  angiosperm:
  - one male gamete fuses with ovum in embryo sac
  - to form diploid zygote
  - other gamete fuses with diploid endosperm cell
  - to form triploid endosperm cell
  - thus double fertilisation takes place

2. (a) **Formation of ovum**
  - formation is inside ovule on upper side of megasporophyll
  - tissue of ovule known as nucellus or megasporangium
  - cells of this tissue are diploid
  - one cell develops to form
  - megaspore mother cell (2n)

- Which divides by meiosis
- to give rise to 4 haploid megaspores (n)
- nucellus can be considered as the female or megasporangium
- 3 of megaspores degenerate
- the fourth divides mitotically to form haploid female gametophyte
- becomes surrounded by diploid, nutritive tissue
- 2 to 3 archegonia develop
- in gametophyte on the side facing the micropyle
- in venter of archegonium develops ovum

(b) **Pollination**
- is pollinated by wind
- pollen grains are dry and light
- air sacs enable pollen grain to remain air-borne
- sporophylls of female cone separate
- pollen is blown between scales and on naked ovules
- adheres to sticky substance
- as it dries, pollen grain is drawn through micropyle
- female cone enlarges
- megasporophylls are firmly cemented by resin

Fertilisation
- occurs one year after pollination
- 2 male gametes move along pollen tube
- towards ovum in archegonium
- tube grows down neck canal
- tip breaks open and male gametes are released
- one fuses with ovum
- to form diploid zygote

Development of seed
- zygote gives rise to embryonic sporophyte plant
- consists of radicle,
- plumule and
- 5 – 8 cotyledons
- an oily endosperm derived from female gametophyte develops
- integument forms the testa
- layer of ovuliferous scale remains attached to testa
- which ultimately forms papery wings
- seeds are dispersed during third year of cone's development
- seeds fall from cone in rotor action to delay fall
- as a result the slightest breeze intercepts and carries it away from parent plant.

4 Invertebrates – Phyla Protozoa and Coelenterata

A. Phylum: Protozoa Example: *Amoeba proteus*

1. Habitat
Amoeba is found in the mud at the bottom of freshwater ponds and slow-flowing streams.

2. Structure
Amoeba is an **unicellular**, **colourless** and **jelly-like** organism. Its shape is **asymmetrical** and it changes its shape constantly.

3. Locomotion
Amoeba moves by means of **pseudopodia** which are extended in the direction of the movement. The **hyaloplasm** (cytosol) flows into the pseudopodium and is withdrawn from the other end so that the organism moves in the direction of the pseudopodium. The **ectoplasm** at the posterior end in the **gel** state, changes into **sol** endoplasm. This **endoplasm** moves toward the anterior end of the pseudopodium where it changes into ectoplasm. This is known as **amoeboid movement**.

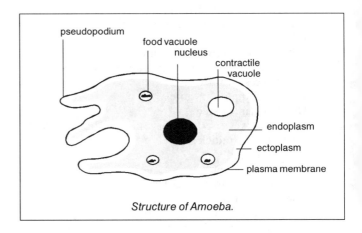

Structure of Amoeba.

4. Nutrition
Amoeba **feeds** on microscopic organisms, e.g. diatoms, desmids, bacteria and decaying plants. It pushes **pseudopodia** out around the food particle and forms a cavity which engulfs it together with a droplet of water in a **food vacuole**. This is known as **ingestion**. Enzymes are secreted by the hyaloplasm which digest the food. Digested food is then absorbed into the endoplasm **(assimilation)**. Undigested parts are left behind by any part of the ectoplasm **(egestion)**.

5. Gaseous exchange
Gaseous exchange takes place by **diffusion** of dissolved oxygen (O_2) from the water environment through the **plasma membrane** into the hyaloplasm and carbon dioxide (CO_2) from the animal to the environment.

6. Reproduction
Reproduction is by **binary fission**. When conditions are **favourable** the nucleus divides by **mitosis** into two parts followed by a division of the hyaloplasm. Each part of the hyaloplasm possesses a nucleus and grows into an adult *Amoeba*. Sometimes under **unfavourable** growth conditions **encystment** takes place. Within the **cyst** the hyaloplasm divides into many parts and each is

surrounded by a **resistant capsule** to form a **spore**. When conditions become **favourable**, the cyst ruptures and each spore develops into a new individual.

7. **Osmoregulation and excretion**
Since *Amoeba* lives in freshwater there is a **higher concentration** of dissolved substances within *Amoeba* than outside. Water thus enters *Amoeba* by **osmosis**. A balance is maintained with the formation of a **contractile vacuole**. The vacuole steadily fills with water, enlarges until it reaches a certain size, bursts and water is removed from the organism in this way. The contractile vacuole then reforms at the same spot. In this way the influx of water is controlled and is known **osmoregulation**. Carbon dioxide (CO_2) and nitrogenous wastes are removed by diffusion through the plasma membrane.

B. Phylum: Coelenterata Example: *Hydra* spp.

1. **Habitat**
The habitat of *Hydra* are **ponds, water-filled ditches** and slow flowing streams where it is attached to submerged objects.

2. **Structure and symmetry**
The body is **cylindrical** and sack-like with a ring of **tentacles** at the **oral** end. The **basal disc** attaches the animal to the substrate. The animal is **radially symmetrical** (if a longitudinal section is made through any diameter, one half will be a mirror image of the other). The body wall is **diploblastic** (two layers of cells), i.e. consisting of an **ecto-** and **endoderm** with in between the **mesoglea**, in which the **nerve network is found. The ectoderm** consists of **musculo-epithelial cells, nematoblasts** (stinging cells) and **sensory cells**. The **endoderm** consists of **nutritive epithelial cells**, i.e. **glandular cells** secreting enzymes, **digestive cells** (amoeboid cells) which can form pseudopodia and **flagellate cells** with flagellae which keep the water in the **coelenteron** in constant motion. Both ecto- and endoderm have **receptor** and **interstitial cells**. The coelenteron is filled with water and is responsible for the animal's **hydrostatic skeleton** (skeleton in which the body fluids themselves provide the structure against which the muscles act).

3. **Locomotion**
Locomotion occurs by a **somersaulting action**, movement of the **pedal disc** and by the secretion of a **bubble** at the basal disc which lets the animal floats.

4. **Nutrition**
Hydra is **carnivorous** and its food consists of **aquatic** (water) animals such as insect larvae, *Daphnia* and other micro-organisms. The prey is caught with **tentacles**, paralysed and/or killed by **nematoblasts** and transported to the single opening to the exterior, situated on the **hypostome**, the **mouth**. **Extra-cellular** digestion takes place in the **coelenteron** by the secretions of the **glandular cells. Flagellate cells** distribute the food and **amoeboid** cells **ingest** it. Food moves by diffusion to the other layers. **Undigested** residues leave the mouth by **egestion**.

5. **Internal transport and gaseous exchange**
Hydra has no special transport system. Its diploblastic nature ensures that all the cells have direct contact with the watery environment. **Gaseous exchange** is directly as a result of **diffusion** from the environment (cf. *Amoeba*).

6. **Reproduction**
Asexual reproduction takes place by means of **budding**. Small *Hydra* grow from the interstitial cells of the ectoderm. Later a **constriction** cuts it off from the parent animal and it develops into a new *Hydra*. Asexual reproduction leads to individuals which are identical to the parents.

   Sexual reproduction takes place with the formation of **ova** (egg cells) in **ovaries** situated near the base of the animal and **sperms** in the **testes**, situated near the tentacles. **Fertilisation** takes place in water (**external**). The **zygote** forms a **cyst** or **theca**, drops off and sinks to the bottom of the pond until conditions improve. The **embryo** then emerges from cyst and develops into a new small *Hydra*. (In some types of *Hydra* a **planula larva** is set free from the cyst).

During **sexual reproduction** the gametes are formed by **meiosis** and thus possess one set of **haploid** chromosomes. With the fusion of gametes from different individuals (**protandrous** in the case of *Hydra*) characteristics are combined which leads to a **larger variety** of desirable traits and ensures greater **variety**. The dormant **encysted stage** ensures **survival** during **unfavourable** conditions.

STANDARD GRADE
QUESTIONS

Section A

A. *Various possibilities are suggested as answers to the following questions. Indicate the correct answer.*

1. *Amoeba* is found in . . . surroundings.
   - A dry
   - B moist
   - C windy
   - D still

2. *Amoeba* is
   - A multicellular.
   - B unicellular.
   - C diploblastic.
   - D triploblastic.

3. A characteristic common to Protozoa is that all of them
   - A form pseudopodia.
   - B are unicellular.
   - C live parasitically.
   - D have contractile vacuoles.

4. Which of the following characteristics does **not** apply to *Amoeba*?
   - A Endoderm
   - B Contractile vacuole
   - C Food vacuole
   - D Ectoplasm

5. Locomotion in *Amoeba* takes place through
   - A crawling.
   - B rolling.
   - C gliding.
   - D pseudopodia.

6. The nutrition of *Amoeba* is described as
   - A autotrophic.
   - B parasitic.
   - C heterotrophic.
   - D holophytic.

7. Ingestion is
   - A the intake of food.
   - B absorption of food.
   - C excretion of undigested food.
   - D digestion of food by enzymes.

8. The process whereby an organism absorbs digested food and makes it part of its own body is called
   - A absorption.
   - B assimilation.
   - C ingestion.
   - D digestion.

9. Which of the following precedes the formation of a food vacuole in *Amoeba*?
   - A Digestion
   - B Cell division
   - C Regeneration
   - D Ingestion

10. Very small animals such as *Amoeba* do not have specially developed breathing organs, because
    A   all cells respire anaerobically.
    B   they are inactive and therefore do not use much oxygen.
    C   they have a large gaseous exchange surface compared to volume.
    D   they are to small to possess special organs.

11. In *Amoeba* the process of mitosis is associated with
    A   fusion.   C   conjugation.
    B   binary fission.   D   osmoregulation.

12. Binary fission in *Amoeba* occurs
    A   asexually.   C   by conjugation.
    B   by spore formation. D   sexually.

13. The structure which is concerned with osmoregulation in *Amoeba* is the
    A   nucleus.   C   contractile vacuole.
    B   food vacuole.   D   plasmalemma.

14. The contractile vacuole, in the diagram below, is numbered

    A   1.
    B   2.
    C   3.
    D   4.

15. The elimination of undigested wastes by *Amoeba* is called
    A   ingestion.   C   assimilation.
    B   excretion.   D   egestion.

16. *Hydra* can be described as
    A   bilaterally symmetrical.   C   diploblastic.
    B   coelomate.   D   triploblastic.

17. The outermost layer of cells in the body wall of Coelenterata is the
    A   ectoplasm.   C   ectodermis.
    B   endoplasm.   D   ectoderm.

18. *Hydra's* symmetry is adapted to
    A   a sedentary existence.   C   a parasitic way of life.
    B   rapid movement.   D   living in water.

19. Which of the following does **not** fit with the rest?
    Locomotion of *Hydra* takes place by means of
    A   amoeboid movement.   C   a gas bubble which lets it float.
    B   somersaulting action.   D   movement of the pedal disc.

20. The skeleton of *Hydra* is
    A   an exoskeleton.   C   an axial skeleton.
    B   an endoskeleton.   D   a hydrostatic skeleton.

21. An organism with intra- as well as extracellular digestion is
    A   *Amoeba*.   C   a reptile.
    B   a mammal.   D   *Hydra*.

22. Which of the following plays an important role in the circulation of small particles of food in the coelenteron of *Hydra*?
    A   Interstitial cells
    B   Flame cells
    C   Flagellate cells
    D   Amoeboid cells

23. In which of the following does digestion of small particles of food take place in *Hydra*?
    A   Interstitial cells
    B   Flagellate cells
    C   Amoeboid cells
    D   Glandular cells

24. Sexual reproduction in *Hydra* takes place by means of
    A   fertilisation.
    B   budding.
    C   constriction.
    D   mitosis.

B. Write down the correct term for each of the following statements.

1. The method of reproduction of *Amoeba*
2. The restoration by an organism of tissue or organs that have been removed or damaged
3. The name of the animal you have studied which reproduces asexual by forming buds
4. The name of the process by which food is taken in by surrounding pseudopodia
5. The method of asexual reproduction in Coelenterata
6. The undifferentiated cells in the body wall of *Hydra*
7. The column of water in the coelenteron of *Hydra* that serves as a type of skeleton
8. The most important function of the contractile vacuole in members of Protozoa
9. Digestion of food inside the cell as in *Amoeba*
10. The structure on which the mouth of *Hydra* is situated

C. The accompanying diagram represents a section through the body wall of Hydra. Name each of the cells A to H in the diagram and next to each the number of the function, from the list below, it is concerned with.

1. Food capture and defense
2. Distribution of food
3. Shortening of the body when it contracts
4. Extracellular digestion
5. Co-ordination of movement
6. Intracellular digestion
7. Detection of stimuli
8. Sexual reproduction and cell replacement

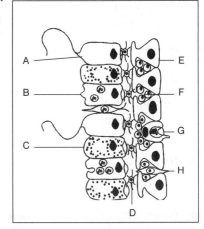

D. Answer the following questions on this animal.

1. Name the phylum to which this animal belongs.
   Give one reason for your answer.
2. Write down the number of the structure concerned with osmoregulation.
3. Which number represents a pseudopodium?
4. What type of digestion occurs in *Amoeba*? Extra- or intracellular?

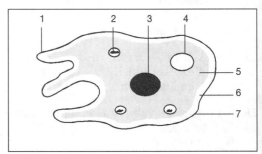

5. Write down the number of the structure concerned with digestion.
6. Write down the number representing the nucleus.
7. Identify the part numbered 7.

E. **Study the diagram of *Hydra* and answer the following questions.**

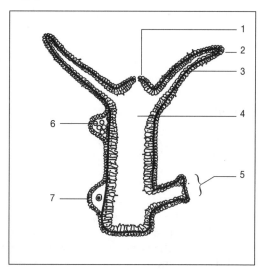

1. Write down the number which represents the coelenteron.
2. What type of digestion takes place in this cavity?
   Extra- or intracellular?
3. What is the disadvantage of having only one opening to the digestive cavity?
4. Identify the structure numbered 5. When is this structure produced in the life cycle of *Hydra*?
5. Write down the number which represents: (*a*) the testis and (*b*) the ovary. By what type of cell division are ova and sperms produced?
6. How does the sperm reach the ovum?
7. What type of symmetry is displayed by *Hydra*?
8. What type of nervous system is found in *Hydra*?
9. What is meant by the term diploblastic?

Section B

1. Describe briefly the structure of *Amoeba*.
2. Describe the locomotion of *Amoeba*.
3. Answer the following questions with regard to the nutrition of *Amoeba*.
   (*a*) Is the mode of nutrition autotrophic, saprophytic or heterotrophic. Give a reason for your answer.
   (*b*) Describe how food is ingested.
   (*c*) Describe what happens to the food in the body until it is assimilated.
   (*d*) What happens to the undigested wastes?
   (*e*) What is the process called by which *Amoeba* gets rid of indigestible residue?
4. Name the different types of nutritive epithelial cells with their functions in the
   (*a*) endoderm, and
   (*b*) ectoderm of *Hydra*.
5. (*a*) What is meant by the term osmoregulation.
   (*b*) Describe the process of osmoregulation in *Amoeba*.
6. What is the process called by which *Amoeba*:
   (*a*) reproduces;
   (*b*) protects itself against unfavourable conditions;
   (*c*) engulfs a food particle together with a droplet of water, and
   (*d*) rids itself of undigested matter?
7. (*a*) Where is *Hydra* usually found?
   (*b*) Describe the structure and symmetry of *Hydra*.
8. Name **three** ways of locomotion in *Hydra*.

9. Answer the following questions on the nutrition of *Hydra:*
   (a) What is the nature of the food?
   (b) How does *Hydra* catches its prey?
   (c) How is the food ingested and digested?

10. Describe gaseous exchange and internal transport in *Hydra*.

11. How does *Hydra* reproduces asexually and sexually?

STANDARD GRADE
ANSWERS

Section A

A. 1. B   2. B   3. B   4. A   5. D   6. C   7. A
    8. B   9. D   10. C   11. B   12. A   13. C   14. C
    15. D   16. C   17. D   18. A   19. A   20. D   21. D
    22. C   23. C   24. A

B. 1. Binary fission   2. Regeneration   3. *Hydra*
    4. Phagocytosis   5. Budding   6. Interstitial cells
    7. Hydrostatic skeleton 8. Osmoregulation   9. Intracellular
    10. Hypostome

C. A Flagellate cell   • 2   B Amoeboid cell   • 6
    C Gland cell   • 4   D Nerve cell   • 5
    E Muscle cell   • 3   F Interstitial cell • 8
    G Nematoblast   • 1   H Sensory cell   • 7

D. 1. Protozoa – body is unicellular   2. No. 4   3. No. 1   4. Intracellular
    5. No. 2   6. No. 3   7. Plasma membrane

E. 1. No. 4   2. Extracellular   3. Unhygenic – for both food intake and excretion of undigested wastes   4. Bud – under favourable growth conditions   5. (*a*)   6. (*b*)   7 – meiosis   6. Swim through the water   7. Radial symmetrical   8. Nerve network   9. Two embryonic layers present only, i.e. ectoderm and endoderm

Section B

1. **Structure of *Amoeba***
    Amoeba is a unicellular, colourless and jelly-like organism; its shape is never the same and changes constantly
    On the outside it is surrounded by a plasma membrane
    The cytoplasma is known as hyaloplasma and consists of a gel ectoplasm and a granular endoplasm in the sol state
    In the endoplasm is the contractile vacuole, food vacuole and nucleus
    Outgrowths, the pseudopodia are used for locomotion and ingestion of food

2. *Amoeba* moves by means of pseudopodia which are extended in the direction of the movement
    Cytoplasm flows into the pseudopodium and is withdrawn from the other end so that the organism moves in the direction of the pseudopodium
    The ectoplasm in the gel state at the posterior end changes into sol endoplasm

This endoplasm moves forward to the anterior end of the pseudopodium where it changes into ectoplasm
This is known as amoeboid movement

3. (a) Heterothrophic: animal cells do not possess chlorophyll; to synthesise their own organic food
   (b) **Intake of food**:
   At any point on surface of body food is taken in
   Pseudopodia develop around the point of contact and a cup-shaped food cavity with a droplet of water is formed
   Pseudopodia meet at tips, fuse and a phagosome is formed
   Process is known as ingestion
   (c) **Digestion of food**:
   Phagosome fuses with lysosome and a secondary lysosome (digestion vacuole) is formed
   Medium at first is acidic and afterwards becomes alkaline; enzymes digest food intracellularly and the process is known as digestion
   Digested food dissolves in the water and diffuses through the membrane of the secondary lysosome into cytoplasm where it becomes part of it, i.e. food is assimilated
   (d) **Getting rid of undigested food**:
   Indigestible remains are left in vacuole, and *Amoeba* rids itself by simply moving away from the residue
   (e) Egestion

4. (a) **Endoderm**:
   Gland cells: secrete enzymes which break food down into smaller particles; extracellularly
   Flagellate cells: cause streaming of water in coelenteron to circulate food
   Amoeboid cells: ingest and digest food intracellularly
   (b) **Ectoderm**:
   Sensory cells: receive stimuli while searching for food
   Nematoblasts: paralyse and entangle prey
   Musculo-epithelial cells: move tentacles in search of food

5. (a) To get rid of surplus water, which constantly enters the body
   (b) **Osmoregulation**:
   Due to osmotically active substances, water enters body by osmosis or taken in with food particles
   The water collects in contractile vacuole
   Energy supplied by ATP from surrounding mitochondria
   Vacuole enlarges, moves to cell surface bursts open to the outside
   A new contractile vacuole forms immediately in its place

6. (a) Binary fission (b) Encystment (c) Ingestion (d) Egestion

7. (a) The habitat of *Hydra* are ponds, water-filled ditches and slow flowing streams where it is attached to submerged objects
   (b) The body of *Hydra* is cylindrical and sack-like with a ring of tentacles at the oral end
   The basal disc attaches the animal to the substrate.
   The animal is radially symmetrical, i.e. if a longitudinal section is made through any diameter, one half will be a mirror image of the other
   The body wall is diploblastic (two layers of cells) i.e. consisting of an ecto- and endoderm with mesoglea in which the nerve network is found, in between
   The ectoderm consists of musculo-epithelial cells, nematoblasts and sensory cells
   The endoderm consists of nutritive epithelial cells, i.e. glandular cells secreting enzymes, digestive cells (amoeboid cells) which can form pseudopodia and flagellate cells with flagellae which keep the water in the coelenteron in constant motion
   Both ecto- and endoderm have receptor and interstitial cells
   The coelenteron is filled with water and is responsible for the animal's hydrostatic skeleton

8. Somersaulting action; gliding movement of basal disc and secretion of gas bubble by basal disc for floating

9. (a) *Hydra* is carnivorous and its food consists of aquatic animals such as insect larvae, *Daphnia,* and other micro-organisms
   (b) The prey is caught with tentacles, paralysed and/or killed by nematoblasts and transported to the single opening to the exterior and situated on the hypostome, the mouth
   (c) Extra-cellular digestion takes place in the coelenteron by the secretions of the glandular cells.
   Flagellate cells distribute the food and amoeboid cells ingest it. Food moves by diffusion to the other layers

10. *Hydra* has no particular transport system; it is diploblastic thus all cells are in contact with the surrounding water
    Each cell on its own exchanges gases and excrete metabolic wastes
    Translocation of food through mesoglea is a result of diffusion

11. **Asexual** by budding
    Small *Hydra* grows from the interstitial cells of the ectoderm
    Later the bud is constricted from the parent animal and develops into a new *Hydra*
    Sexual reproduction occurs by the formation of egg cells in ovaries near the base of the animal and sperms in testes, in the ectoderm, near the tentacles
    Fertilisation is external in the water; the zygote forms a cyst or theca, falls of and lies on the bottom of the pool until conditions improve; the embryo emerges from the cyst and develops into a new *Hydra* when growth conditions improve

HIGHER GRADE
QUESTIONS

Section A

A. *Various possibilities are suggested as answers to the following questions. Indicate the correct answer.*

1. Which statement concerning Protozoa is **incorrect**? They
   A live mainly heterotrophically.
   B are free-living, but some are parasitic.
   C move by means of cilia or pseudopodia.
   D survive unfavourable conditions by aestivation.

2. Which of the following is **not** relevant to *Amoeba?*
   A Ectoplasm   C Contractile vacuole
   B Ectoderm   D Food vacuole

3. *Amoeba* which is moved from water into a weak sucrose solution would be predicted to
   A show increased activity of its contractile vacuole.
   B show decreased activity of its contractile vacuole.
   C show increased water potential of its endoplasm.
   D absorb sucrose from its surroudings.

4. All *Amoeba* spp. are
   A parasites.   C heterotrophs.
   B saprophytes.   D decomposers.

5. *Amoeba* and bacteria are similar in that both have
   - A cell walls.
   - B contractile vacuoles.
   - C a distinct nucleus.
   - D binary fission.

6. Osmoregulation is related to
   - A intracellular digestion.
   - B excretion.
   - C reproduction.
   - D ingestion.

7. An *Amoeba* spp. encysts
   - A to reproduce.
   - B to hide itself from attack by predators.
   - C to withstand unfavourable conditions.
   - D before binary fission.

8. Which of the following sets refers to the Coelenterata?
   - A Acellular
   - B Multicellular, diploblastic, tissue level of organisation
   - C Multicellular, diploblastic, organ level of organisation
   - D Multicellular, triploblastic, tissue level of organisation

9. The nerve network of *Hydra* is found in the
   - A ectoderm.
   - B endoderm.
   - C mesoglea and ectoderm.
   - D mesoglea.

10. The relationship of some *Hydra* spp. with zoochlorella is an example of
    - A parasitism.
    - B mutualism.
    - C commensalism.
    - D saprophytism.

11. *Hydra* may move about by somersaulting. In the diagram, it has just attached its tentacles to a stone and is about to raise its foot. To do this the muscle cells in the ectoderm at X must now
    - A relax while those at Y contract.
    - B contract at the same time as those at Y.
    - C relax at the same time as those at Y.
    - D contract, while those at Y relax.

12. The most important method of protection in *Hydra* is to
    - A use its tentacles to ward off the enemy.
    - B contract body and tentacles.
    - C flee from the enemy.
    - D paralyse its enemy.

13. Digestion in *Hydra* occurs
    - A intracellularly only.
    - B extracellularly only.
    - C both intra- or extracellularly.
    - D internally near the tentacles.

14. Nematoblasts are triggered
    - A by any change in the water around the animal.
    - B when the prey gets close enough.
    - C by chemicals touching the cnidocil.
    - D by stimulating the nerve network as a result of the action of sensory cells.

15. The male organs of *Hydra* develop
    - A externally near the tentacles.
    - B internally near the hypostome.
    - C externally near the base.
    - D internally near the tentacles.

B. Write down the correct term for each of the following statements.

1. Body plan where similar structures are arranged around a central axis
2. The temporary, wrinkled posterior end of *Amoeba*
3. Locomotion response to gravity
4. Ingestion of liquid droplets in a vesicle through the cell membrane
5. Uptake of solid particles in a vesicle and into a cell
6. Male and female organs on the same individual
7. The free-living ciliated embryo of certain Coelenterata

Section B

1. Give the genus of the member of Protozoa you have studied. Describe why it is considered to be
   (*a*) a living organism, and
   (*b*) an animal.

2. The following diagram represents a longitudinal section through the body wall of *Hydra*.

   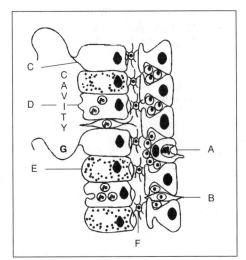

   (*a*) Identify the cell numbered A. Explain how it assists the organism in capturing food.
   (*b*) Explain the role played by cell types C, D and E in the digestion of food
   (*c*) In the organism name the space numbered G. What does this space normally contain?
   (*d*) Write down the letter which represents the nervous system in the diagram. What type of nervous system is it?

3. (*a*) Give a definition of the term "sexual reproduction".
   (*b*) Explain the terms:
      (i) gamete, and
      (ii) zygote.
   (*c*) Some species of *Hydra* are hermaphroditic. Animals which are stationary are often hermaphroditic while those move about usually have separate sexes.
      (i) How do a sessile animal benefit from being hermaphroditic?
      (ii) What is the disadvantage of an animal being hermaphroditic?

4. Describe the concepts:
   (*a*) osmoregulation, and
   (*b*) excretion.

5. Study the following diagrams of *Amoeba* and describe the processes of osmoregulation and excretion.

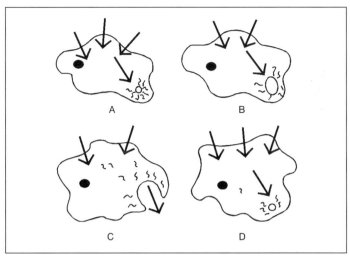

6. Contractile vacuoles occur in most Protozoa. Large groups of *Amoeba* were kept in different cultures of diluted sea water and the rate of elimination of water by their contractile vacuoles was measured. The result are shown in the table below.

% of sea water (concentration of medium)	10	15	20	25	30
Average fluid elimination (parts of water/million *Amoeba*/minute)	8	5	4	3	1

(a) Plot a graph by using the data above.
(b) The contractile vacuole serves two functions in *Amoeba*. It excretes nitrogenous wastes and/or remove excess water. Which of these two functions is supported by the results above? Explain your answer
(c) Explain the following observations:
   (i) The rate of contraction of the vacuole often increases when the animal feeds.
   (ii) Electron micrographs reveal a ring of mitochondria around the contractile vacuole.

7. "*The development of a nervous system of an animal is directly proportional to the advancement of the animal*". Briefly discuss this statement referring to nervous co-ordination in *Hydra*.

8. Explain the significance of:
   (a) the microscopic unicellular state to *Amoeba*, and
   (b) radial symmetry to *Hydra*.

9. Describe the sexual reproduction of *Hydra*.

Section C

Compare the nutrition of *Hydra* and *Amoeba* with reference to the following:
(a) how they obtain their food,
(b) intake of the food, and
(c) digestion of food.

HIGHER GRADE
ANSWERS

Section A

A. 1. D   2. B   3. B   4. C   5. D   6. B   7. C
    8. B   9. D   10. B   11. D   12. B   13. C   14. C
    15. A

B. 1. Radial symmetrical   2. Uriod   3. Geotaxis
    4. Pinocytosis   5. Phagocytosis   6. Hermaphrodite
    7. Planula

Section B

1. *Amoeba*
   (a) **Nutrition:** food is ingested in a solid form
       Reproduction: asexual takes place by binary fission
       Respiration: takes up oxygen (O_2) and gives off carbon dioxide (CO_2)
       Growth: when maximum size is reached it divides into two
       Excretion: takes place by diffusion from the cell
       Irritable: moves towards poor light and away from harmful substances
       Locomotion: moves freely with pseudopodiums
   (b) *Amoeba* feeds holozoically, there are no plastids
       The body wall has no cell wall, only a plasma membrane
       It moves freely with pseudopodia or cilia

2. (a) Nematoblast
       The tentacles execute swaying movements and when water flea/insect touches the cnidocil, the nematoblast expels a nematocyst with coiled tubule
       The penetrants paralyse the prey and the volvents entangle the prey
       They are found on the tentacles
   (b) C: flagellate cell keeps water in coelenteron in circulation
       D: amoeboid cell ingest the food
       E: glandular cell secrete enzymes for digestion of food
   (c) Coelenteron; water
   (d) F: nerve network

3. (a) **Sexual reproduction**
       Any kind of reproduction involving the fusion of gametes to form a zygote
   (b) (i) Gametes are haploid cells which fuse to form a zygote
       (ii) The zygote is a cell formed by fusion of two gametes or reproductive cells
   (c) (i) It has little chance of meeting sexually mature partner of its own kind; if not fertilised by another individual self-fertilisation can take place
       (ii) A disadvantage is the possibility of self-fertilisation taking place

4. (a) **Osmoregulation**
       This is the maintenance of a constant osmotic pressure and comprises mainly problems of water and ion balance
       The thin plasma membranes around cells create osmotic problems because water moves along a concentration gradient

The cells could either shrink or burst depending on the outside concentration
Ions tend to equalise concentration inside and outside the cells
The cell membrane can transport substances actively, i.e. against the concentration gradient
(b) **Excretion**
Nitrogenous wastes are formed during metabolism
These substances, such as urea and ammonia, are toxic when accumulated in cells
These wastes, excess carbon dioxide (CO_2) and water (H_2O) are removed and excreted in a dissolved form

5. Metabolism produces wastes products, e.g. urea, carbon dioxide and water
Dissolved wastes present in high concentrations in the hyaloplasm creates a high osmotic pressure inside the cell
The continuous inflow of water by osmosis across the differentially permeable membrane, the plasmalemma, causes the contractile vacuole to act as a pump to eliminate excess water
Water is absorbed into the contractile vacuole as fast as it enters the cell
This is an active process requiring energy supplied by ATP from the mitochondria
The contractile vacuole enlarges as water enters and when it reaches maximum size water is discharged through a pore and the osmotic pressure is thus preserved
A new contractile vacuole is formed immediately in its place and the process is repeated
Nitrogenous wastes move through the plasmalemma and is excreted in a solid form
Carbon dioxide diffuses outwards, from a high to a low concentration, into the surrounding water

6. (a)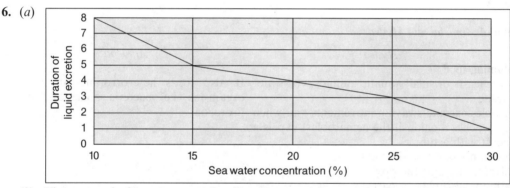

(b) The removal of excess water. As the sea water concentration increases, the water diffuses out of *Amoeba's* hyaloplasm. The water concentration inside *Amoeba* respectively decreases and less water is excreted by the contractile vacuole
(c) (i) Water is taken into the body with the food in the food vacuole. This water has to be eliminated and the rate of contraction increases
(ii) Contraction of the vacuole requires energy which is provided by the mitochondria

7. **Nervous co-ordination in *Hydra***
The degree of development of the nervous system is related to its complexity
Hydra has a poorly developed nerve network over the entire body
It consists of undifferentiated nerve cells in the mesoglea and sensory cells in the ecto- and endoderm. Neurons are connected to the musculo-epithelial cells and when touched these cells respond by contracting into a spherical mass
Sensory cells receive stimuli and convert it into nerve impulses which are transmitted to neurons and then to the musculo-epithelial cells. This makes reaction to stimuli possible
A central nervous system is absent and impulses are received and transmitted in any direction which is an adaptation to *Hydra's* radial symmetry

8. (a) **Unicellular state**
The ratio of surface area to volume is large and all parts of body are able to exchange materials directly with the environment

Life is thus maintained without diversity of highly specialised structures
Its minuteness makes them vulnerable to the environment

(b) **Radial symmetry:**
There is no dorsal or ventral side, but only an oral and an aboral side
They are mostly sessile (stationary)
A whorl of tentacles allows the animal to sense and encounter food and enemies equally well from all directions

9. **Sexual reproduction of** *Hydra*
Sexual reproduction takes place in the summer or spring
The animal is usually bisexual and the testes develop as buds just below the tentacles by division of the interstitial cells
Meiosis takes place and sperms are formed inside the testes, it burst open and sperms are released in the water
Ova (egg cells) are formed from interstitial cells in ovaries situated near the base of the animal
One cell divides by meiosis and forms the ovum
The ectoderm bursts and the ovum is exposed to the water
Fertilisation takes place when the nucleus of one sperm fuses with the nucleus of the egg cell to form a zygote

Section C

Nutrition of *Amoeba* **and** *Hydra*

Hydra	*Amoeba*
(a) **Obtaining food**	
Its food consists of small aquatic animals, viz, water fleas, larvae	It also feeds heterotrophic on aquatic organisms, i.e. bacteria algae and other protozoa
The tentacles execute swaying movements and when the prey touches a cnidosil the nematoblast expels a nematocyst with a tubule	
The penetrants paralyse the prey	
The volvents entangle the prey	
Gluthathione also plays a part	
(b) **Intake of food**	
The immobilised prey is transferred to the hypostome and drawn into the mouth by muscular contractions of the musculo epithelial cells into the coelenteron	When food is encountered the pseudopodia forms around the food forms a food cavity Together with a small amount of water it forms a food vacuole (phagosome) and is known as ingestion
The cell of the hypostomal region secrete mucus which assists in swallowing. Currents of water in the coelenteron also assist with this process.	It can occur at any point on the plasmalemma and is an example of phagocytosis
(c) **Digestion of food**	
The glandular cells secrete enzymes which digest the food partially in the coelenteron, i.e. extracellular digestion	Digestion takes place and the enzymes synthesised on the ribosomes move through the ER to the Golgi apparatus and is surrounded with a membrane to form a lysosome
Flagellate cells cause currents in the water of the coelenteron and in this way the food is circulated	
The amoeboid cells ingest organic parts by phagocytosis and within the food vacuole the enzymes digest the food intracellularly	The medium at first is acid but later it becomes alkaline
The translocation of food is by diffusion through the mesoglea to the ectoderm	The digested food is assimilated into the cytoplasm and digestion is intracellular

5 Invertebrates – Phyla Platyhelminthes and Annelida

A. Phylum; Platyhelminthes Example: *Taenia solium* – the Tapeworm

1. Habitat
The tapeworm is **parasitic** and a **multicellular** animal. The adult stage lives in the small intestine of man where it is attached to the **intestinal wall** by means of **hooks** and **suckers**. Its body lies free in the intestinal canal and is bathed in digested food of the **host**.

2. Structure
The adult worm is approximately 5 m long and 6 mm wide. The body is **flat, bilateral symmetrical** (the condition in which one half of the organism, from a section taken down its long axis, is a mirror image of the other half) and tapers to the anterior (front) end. The **head** or **scolex** is small with two rows of **hooks** and four **suckers**. The **neck** is the part where young **proglottides** (segments) are formed by **strobilation**. Older segments occur at the posterior (hind) end and the younger ones, in which the **gonads** (reproductive organs) are not yet formed, at the neck. In the **mature proglottides**, far from the neck, the **male** and **female gonads** are fully developed. The older proglottides at the tip are filled with **eggs** and the male gonads have degenerated. **Digestive** and **respiratory systems** are absent and the **excretory** and **nervous systems** are simple.

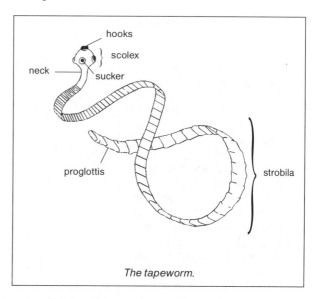

The tapeworm.

3. Nutrition
Nutrition takes place by **diffusion** of digested food from the intestine of the host. Its **flattened shape** supplies the **largest** possible **area to volume ratio** for the absorption of food. Its **flattened form** ensures that all cells are close to the ectoderm or parts of the alimentary canal. Food diffuses from the alimentary canal to all parts of the body.

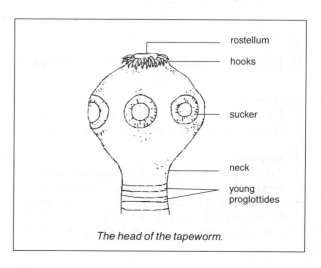

The head of the tapeworm.

4. **Gaseous exchange and internal transport**

   Gaseous exchange takes place by **diffusion** of dissolved oxygen from the watery environment into the ectodermal cells and from there to the mesoderm. Outward cell to cell diffusion of carbon dioxide also occurs.

   Excretory products and water diffuse from the cells with the aid of **flame cells** into small tubles.

5. **Reproduction and life-cycle**

   The worms are **bisexual** and **self-fertilisation** take place. A **secondary host** is required. The tapeworm has proglottides in various stages of development, in the intestine of man, the **host**. The male gonads develop and reach maturity first. Each mature proglottide (at the posterior end) has both male and female gonads. The male gonads develop and ripen first. The younger proglottides (near the neck) possess no gonads and are followed by those possessing only male organs. Those with both male and female organs follow and finally those in which the male gonads have degenerated and in which only **fertilised eggs** occur.

   The older proglottides at the posterior end of the worm drop off and are passed of with the **faeces** of man (host). The **hexacanth** develops within the egg. This embryo is surrounded by a **hard shell** so that it can survive in the soil and develops no further unless it is ingested by a **pig (secondary host)**. The digestive juices of the pig's stomach dissolve the shell around the embryo. The embryo then **bores** with its **hooks** and suckers through the **intestinal wall** into a **blood vessel** and is transported to a **muscle** where it **encysts** to form a **cysticercus** (bladderworm). It develops no further in the pig. If **poorly cooked** pork is eaten it can develop further. In the intestine of man the **proscolex** everts from the bladder and attaches itself to the intestinal wall. New proglottides start to develop and the life-cycle starts all over again.

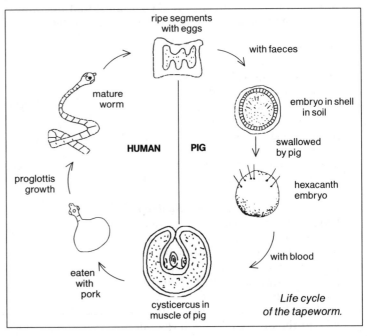

Life cycle of the tapeworm.

B. Phylum: Annelida Example: *Lumbricus terrestris* – the Earthworm

1. **Habitat**

   The earthworm is a **terrestrial** animal and occurs in **moist soil** through which it **tunnels**.

2. **Structure**

   The body is **segmented** and is **bilaterally symmetrical**. The advantage of this, together with **cephalisation** (the tendency of sensory organs to be situated in the head or head region), is that it can survey the environment better as it moves

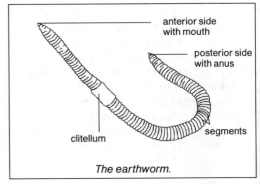

The earthworm.

forward. The disadvantage is that the body cannot move efficiently in any direction. It also cannot receive stimuli efficiently from all directions.

The **epidermis** is covered with a thin **cuticle** and the animal is **triploblastic**. A **coelom** occurs in the **mesoderm**. The **nerve cord** is situated ventrally. In each segment is a pair of excretory organs, the **nephridia**. A thickened part, the **clitellum**, occurs on segments 32 to 37.

The **mouth** is situated behind the **prostomium** on segment one and the **anus** on the last segment. **Nephridiopores** occur on all the segments except the first three and the last one. The **male sexual openings** (one pair) open ventral on segment 15 and the **female openings** (one pair) ventral on segment 14. From the ninth segment onwards are the **dorsal pores**. On all segments except the first and last segment there are four pairs of **chaetae**.

The body is **pliable** (to tunnel in the soil) and the liquid in the coelom acts as a **hydrostatic skeleton**.

3. **Locomotion**

The earthworm **moves** forward by **muscular contractions** that alternately shorten and lengthen the body. By anchoring the chaetae at the anterior part of the body and shortening the body (contraction of the longitudinal muscles) the posterior end is pulled forward. The anterior end is now anchored by chaetae, the body lengthens (contraction of circular muscles) and the anterior end is pushed forward.

4. **Nutrition**

Earthworms **feed** on decaying **organic material** in the soil. By **ingestion** food is taken in through the mouth on segment one. The **through gut** makes a continuous digesting and **specialising** process possible in the various parts. From the **mouth** food moves through the **pharynx** where it is mixed with **mucus** and **enzymes**. The **oesophagus** secretes more enzymes and **calcium carbonate** ($CaCO_3$) by the **calciferous glands**. In the **crop** more enzymes are secreted and in the **gizzard** food are thoroughly mixed with the enzymes and finely **grounded**. **Digestion** and **absorption** take place in the small intestine. Digestion is thus **extracellular**. The blood transports the digested food to all parts of the body. The absorption surface is enlarged by the **typhlosole** and **long alimentary canal**. **Egestion** of undigested wastes takes place through the **anus** on the last segment. The **advantages** of a through-gut are that digested food is never mixed with outgoing waste; different parts make digestion more efficient; food can be kept for a considerable time until digestion is completed and large quantities of food can be ingested.

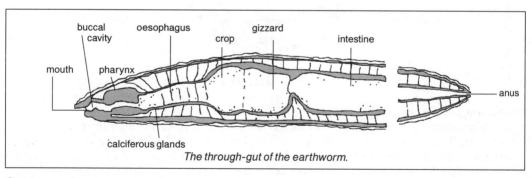

The through-gut of the earthworm.

5. **Gaseous exchange**

Gaseous exchange takes place by **diffusion** of oxygen through the **moist** body surface. The skin is kept moist by **mucus glands** and a **moist environment**. From the epidermal cells the oxygen diffuses into the **blood** and is transported by the pigment **erythrocruorin**. Carbon dioxide diffuses out from the blood to the **soil air**.

6. **Internal transport** (HG only)
   The blood system is a **closed one** and consists of **dorsal**, **ventral** and **subneural** blood vessels. **Pseudohearts** create a circulation by contracting and join the dorsal and ventral blood vessels.

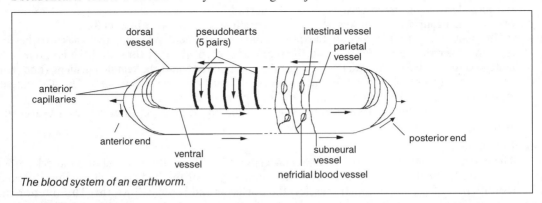
The blood system of an earthworm.

Parietal blood vessels in each segment join the dorsal and **subneural** blood vessels. The **intestinal blood vessels** supply the small intestine and the **nephridial blood vessels** the nephridia. The function of blood is the **distribution** of **respiratory gases** as well as **food** and the **transport** of **waste products** to the excretory organs.

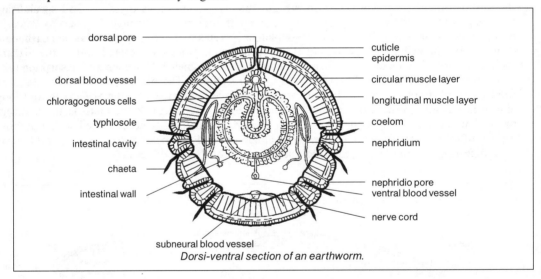
Dorsi-ventral section of an earthworm.

7. **Osmoregulation and excretion**
   Osmoregulation and excretion of nitrogenous wastes take place through the two nephridia per segment. It consists of a **nephrostome** in the coelom and a **nephridiopore** found on all segments except the first three and last segment. Wastes from the coelom move into the nephrostome and out through the pore. **Useful substances** and **water** are reabsorbed from the tube back into the blood.

8. **Reproduction**
   The earthworm is **hermaphroditic**. During **copulation** an exchange of sperm takes place that is stored in **spermathecae**. A mucus sheath prevents the sperm from mixing. The openings of the spermathecae are between segments 9 and 10, and 10 and 11. Later a **cocoon** with albumin food in it is formed by the **clitellum**. Sperm and eggs are released into it and fertilisation is **external** inside the **cocoon**. The young worms develop in the cocoon and feed on **albumin**.

   An advantage of hermaphroditism is that during copulation **both** individuals are **fertilised**. Chances of survival are enhanced. A disadvantage is that self-fertilisation is possible.

STANDARD GRADE
QUESTIONS

Section A

A. *Various possiblities are suggested as answers to the following questions. Indicate the correct answer.*

1. Which of the following statements concerning Platyhelminthes is incorrect? They
   A   are triploblastic.
   B   sometimes live parasitic.
   C   possess a well developed muscular system.
   D   possesses no coelom.

2. All members of Platyhelminthes are
   A   round as seen in transverse section.
   B   flattened laterally.
   C   flattened dorso-ventrally.
   D   radially symmetrical.

3. The tapeworm gets food by
   A   producing food autrotrophically.
   B   synthesising inorganic food.
   C   diffusion from the environment.
   D   taking in organic food through its mouth.

4. The gaseous exchange surface of the tapeworm is its
   A   body surface.
   B   gills.
   C   lungs.
   D   mouth.

5. Fertilisation in the tapeworm takes place in the
   A   small intestine.
   B   large intestine.
   C   oviducts.
   D   genital atrium.

6. The tapeworm is attached to the human intestinal wall by means of
   A   hooks only.
   B   suckers only.
   C   hooks and suckers.
   D   its mouth.

7. The tapeworm gives rise to new proglottides by means of
   A   meiosis.
   B   strobilation.
   C   regeneration.
   D   cephalisation.

8. The protective layer of the ripe proglottides of the tapeworm is known as
   A   an integument.
   B   a pleuron.
   D   a tergum.
   C   a tegument.

9. The zygote of *Taenia solium* develops into a larvae with six hooks, called a/an
   A   embryophore.
   B   onchosphere.
   C   hexacanth.
   D   cysticercus.

10. The intermediate host of the tapeworm is
    A   man.
    B   the pig.
    C   a fish.
    D   the earthworm.

11. This structure is a/an
    A   proglottide.
    B   cysticercus.
    C   proscolex.
    D   everted cysticercus.

12. The habitat of the earthworm is
    A   dry soil.
    B   gravel.
    C   waterlogged soil.
    D   moist soil.

13. The symmetry of the earthworm can be described as
    A   radial.
    B   bilateral.
    C   triploblastic.
    D   diploblastic.

14. Which of the following does **not** fit? The earthworm has
    A   an epidermis covered with a cuticle.
    B   a coelom situated in the mesoderm.
    C   ventrally situated nerve string.
    D   a pair of nephridia in each segment.

15. The coelom of an earthworm
    A   separates the ectoderm from the endoderm.
    B   digest the food.
    C   contains a fluid.
    D   separates mesoderm from ectoderm.

16. Which of the following causes and earthworm to become shorter and thicker?
    A   Loss of water through the nephridiopore
    B   Elongation of the alimentary canal
    C   Contraction of the circular muscles
    D   Contraction of the longitudinal muscles

17. The sensitive, small structure found in front of the mouth opening of *Lumbricus* is the
    A   clitellum.
    B   peristomium.
    D   typhlosole.
    C   prostomium.

18. The absorption surface area of the intestine in Annelida is increased by the
    A   typhlosole.
    B   nephridia.
    C   nephrostome.
    D   flame cells.

19. The typhlosole is of importance to the earthworm because it
    A   increase the absorption surface area of the intestine.
    B   is the essential part of the respiratory system.
    C   contains important sex organs.
    D   makes the dorsal body wall more firm.

20. Gaseous exchange in the earthworm
    A   takes place through the mouth.
    B   is dependent on the transport of oxygen by haemoglobin.
    C   takes place between the environment and the blood vessels in the body surface.
    D   is absent because the worm moves so slowly.

21. The dorsal pore of the earthworm is mainly concerned with
    A   nutrition.
    B   the absorption and storage of sperm during copulation.
    C   keeping the body surface moist.
    D   locomotion.

22. Nephridia are the excretory organs of
    A   Protozoa.
    B   Coelenterata.
    C   Platyhelminthes.
    D   Annelida.

23. The structure responsible for osmoregulation in *Lumbricus*, is the
    A   clitellum.
    B   flame cells.
    C   typhlosole.
    D   nephridia.

24. How many pairs of nephridiopores are present in an earthworm with 180 segments?
    A   180
    B   178
    C   176
    D   90

25. The structures which are responsible for the excretion of liquid wastes from organismes are
    A   nephridia and food vacuoles.
    B   nephridia and contractile vacuoles.
    C   contractile vacuoles and pseudopodia.
    D   nephridia and anus.

26. During copulation between two earthworms male gametes are transferred
   A   to the spermathecae of each worm.
   B   always from young earthworm to the older one.
   C   to the ovary of each worm.
   D   to the mucous sheath of each worm.

B. Write down the correct term for each of the following statements.

1. The front part of the body of the tapeworm with which it attaches itself to the wall of the intestine
2. The inverted scolex of the cysticercus
3. Respiration without molecular oxygen as in *Taenia solium*
4. The tendency of sensory organs to be situated in the head or head region
5. The repetition of similar segments of which the bodies of Annelida are composed
6. The fold in the dorsal wall of the alimentary canal of the earthworm
7. The sensitive appendage on the first segment of an earthworm
8. The excretory organs concerned with osmoregulation in the earthworm
9. The organs in the earthworm in which sperm of another worm is temporarily stored
10. The structures of the earthworm with which it anchors itself in the soil
11. The thickening from segments 32 to 37 in the earthworm

C. The diagram below represents a transverse section through an earthworm.

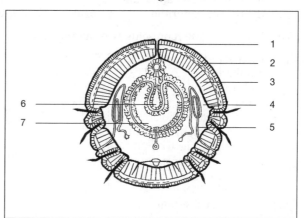

1. Write down the number of the parts concerned with osmoregulation. State one other function of this organ.

2. Identify the fold numbered 6. What is the function?

3. Identify the space numbered 4. With what is this space filled?

4. Write down the numbers of two muscle layers shown in the diagram and state the function of each.

D. The following diagrams represent stages in the life cycle of the tapeworm. Study it and answer the following questions.

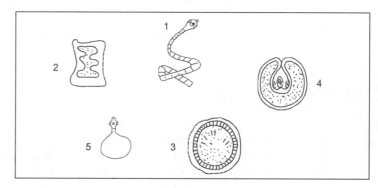

1. Identify the stages numbered 1 to 5 and indicate where each is found.

2. How would one protect oneself against tapeworm infection?

Section B

1. (a) Where is the adult tapeworm found?
   (b) Briefly describe the structure of the adult tapeworm.

2. (a) Where and how does the tapeworm obtain its food?
   (b) How are the tapeworm adapted for the absorption of food?

3. Briefly describe
   (a) gaseous exchange; and
   (b) internal transport in the tapeworm.

4. (a) Mention the name of the example of Platyhelminthes you have studied.
   (b) How does gaseous exchange takes place in this animal?
   (c) How is the body of this animal adapted to this type of gaseous exchange?
   (d) These animals possess no blood system. How does food reach the various parts of the body?
   (e) How do undigested food residues leave the body?

5. Briefly describe the (a) habitat, and (b) the structure of the earthworm.

6. Briefly describe the mechanisms of locomotion in the earthworm.

7. Name, in the correct order, the different parts of the alimentary canal of the earthworm from the mouth to the anus. Write next to each the function of the particular part.

8. Describe:
   (a) the gaseous exchange, and
   (b) the role of nephridia in osmoregulation and excretion in the earthworm.

9. The diagram below represents two mating earthworms. Answer the following questions.

   (a) Identify the openings numbered 1, 2 and 3, and the structure numbered 4.
   (b) What prevents the sperm from A being mixed with the sperm from B?
   (c) How is the sperm transported from worm A to worm B?
   (d) Actual fertilisation only takes place later. Is the fertilisation process of the earthworm external or internal?
   (e) What is the function of the clitellum?

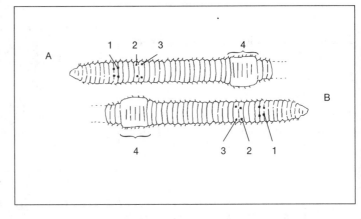

STANDARD GRADE
ANSWERS

Section A

A. 1. C   2. C   3. C   4. A   5. C   6. C   7. B
   8. D   9. C   10. B   11. D   12. D   13. B   14. D
   15. C   16. D   17. D   18. A   19. A   20. C   21. C
   22. D   23. D   24. C   25. B   26. A

B. 1. Scolex   2. Proscolex   3. Anaerobic
       4. Cephalisation   5. Metameres   6. Typhlosole
       7. Prostomium   8. Nephridia   9. Spermathecae
      10. Chaetae   11. Clitellum

C. 1. No. 5; excretion of metabolic waste
       2. Typhlosole; it increases the absorption surface for food in the earthworm
       3. Coelom; filled with coelomic fluid
       4. No. 2; circular muscle layer; contraction lenghtens the body of the earthworm
          No. 3; longitudinal layer; contraction shortens the body of the earthworm

D. (*a*) 1. Adult tape worm – alimetary canal of man
            2. Mature proglottide – in soil
            3. Onchosphere – in soil
            4. Cysticercus (bladderworm) – in muscles of pig
            5. Cysticercus with everted scolex – alimentary canal of man
       (*b*) Avoid contact with untreated sewage
            Make sure that pork is free of measles when eaten
            Meat should be thoroughly cooked

Section B

1. (*a*) In the alimentary canal of man
   (*b*) The adult worm is approximately 5 m long and 6 mm wide
        The body is flat, bilateral symmetrical and tapers to the anterior end
        The head or scolex is small; with two rows of hooks and four suckers
        The neck is the part where young proglottides (segments) are formed; by stobilation
        Older segments occur at the posterior end; and the younger ones, in which the gonads are not yet formed, at the neck
        In the mature proglottide, far from the neck, the male and female gonads are fully developed
        The older proglottides at the tip; are filled with eggs; and the male gonads have degenerated
        Digestive and respiratory systems are absent and the excretory and nervous systems are simple

2. (*a*) Nutrition takes place by diffusion; of digested food from the small intestine of man
   (*b*) It flattened form; supplies the largest possible surface area; to volume ratio; for the absorption of food

3. (*a*) Oxygen diffuses from the watery environment; into ectoderm cells; and then from cell to cell into mesoderm
        Carbon dioxide; diffuses from cell to cell outwards
   (*b*) Transport of food and gases takes place by diffusion
        Its flattened form ensures that all cells are in close contact with the ectoderm

4. (*a*) *Taenia solium* or the tapeworm
   (*b*) Diffusion; through its body surface
   (*c*) The body is flattened; and all cells are near the external environment
   (*d*) Dissolved food diffuses from cell to cell
        Its flattened body ensures that all cells are near food
   (*e*) The food that is absorbed is already digested; so there is no undigested wastes

5. (a) The earthworm is terrestrial animal; and are found in moist soil through which it tunnels
   (b) The body is segmented and is bilaterally symmetrical
   The epidermis is covered with a thin cuticle and the animal is triploblastic
   A coelom occurs in the mesoderm and the nerve cord is situated ventrally
   In each segment a pair of excretory organs
   A thickened part, the clitellum, occurs on segments 32 to 37
   The mouth is situated behind the prostomium on segment one and the anus on the last segment
   Nephridiopores occur on all the segments; except the first three and the last one.
   The male sexual openings (one pair) open ventral on segment 15 and the female openings (one pair) ventral on segment 14
   From the ninth segment onwards are the dorsal pores
   On all segments; except the first and last segment; there are four pairs of chaetae.
   The body is pliable; and the liquid in the coelom acts as a hydrostatic skeleton.

6. The earthworm moves forward by muscular contractions; which alternately shorten and lengthen the body
   By anchoring the chaetae; at the anterior part of the body; and shortening the body by contraction of the longitudinal muscles; the posterior end is pulled forward
   The anterior end is now anchored by chaetae; and the body lengthened; by contraction of circular muscles; and the anterior end is pushed forward

7. Pharynx; food mixes with mucus and enzymes
   Oesophagus; enzymes secreted; calciferous glands excrete lime
   Crop; more enzymes secreted
   Gizzard; food mixed with enzymes and ground to a fine pulp
   Intestine; absorption of digested food
   Anus; excretion of faeces

8. (a) Gaseous exchange occurs through the body surface
   Oxygen dissolves in moisture; on body surface; and diffuses into blood capillaries to be transported to the tissues
   Carbon dioxide is released in the same way
   (b) The flickering movements of cilia in nephrostome; drive coelomic fluid forward in coiled tube
   It contains nitrogenous wastes; and as it moves along the tube capillaries reabsorb water and useful substances. Only wastes remain to be excreted
   Wastes are excreted through nephridiopores to outside; when sphincter muscle relaxes

9. (a) 1 – openings of spermathecae
      2 – opening of oviduct
      3 – opening of sperm duct
      4 – clitellum
   (b) Each is enclosed in its own mucus sheath
   (c) Along seminal groove in epidermis
   (d) External
   (e) It secretes a membrane which will form the cocoon for zygotes; and later for the embryo's
      Albumen is secreted as food for the young
      During copulation the gland cells of the clitellum secretes a common mucus sheath

HIGHER GRADE
QUESTIONS

Section A

A. *Various possiblities are suggested as answers to the following questions. Indicate the correct answer.*

1. *Taenia solium* is an example of an animal which
   A   is radially symmetrical with an open blood system.
   B   possesses no alimentary and transport system.
   C   possesses nephridia which are responsible for osmoregulation.
   D   digests food intracellularly and circulate it in an coelenteron.

2. Osmoregulation is related to
   A   ingestion.   C   intracellular digestion.
   B   excretion.   D   gaseous exchange.

3. Platyhelminthes are termed acoelomate because they
   A   only have one external opening.
   B   are bilaterally symmetrical.
   C   possess a body cavity.
   D   have no cavity in the mesoderm between the ectoderm and the endoderm.

4. Cephalisation is most likely to be associated with
   A   a primitive nerve network.   C   sedentary types of animals.
   B   radial symmetry.   D   bilateral symmetry.

5. The type of gut shown in this diagram of an animal is
   A   characteristic of Phatyhelminthes.
   B   characteristic of Coelenterata.
   C   found only in man.
   D   a through-gut.

6. Which of these is **not** true concerning the gaseous exchange surface of *Taenia solium?*
   A   Moist   C   Well supplied with blood
   B   Possesses a thick cuticle   D   Gases diffuse through in solution

7. Earthworms feed mainly on
   A   smaller earthworms.   C   rotting plant material in the soil.
   B   young leaves of seedlings.   D   the cell sap of roots.

8. The hydrostatic skeleton of the earthworm is the
   A   blood in the ventral blood vessel.   C   blood in the dorsal blood vessel.
   B   digestive juices in the gut.   D   liquid in the coelom.

9. The nephridiopores are present in an earthworm
   A   in the grooves between segments 9 and 10, and 10 and 11.
   B   on all the segments.
   C   on all the segments except the first three.
   D   on all the segments except the first three and the last one.

10. Earthworms are adapted for gaseous exchange in that
    A   the typhlosole enlarges the absorption surface area.
    B   the blood contains erythrocruorin.
    C   the skin is moist and richly supplied with blood capillaries.
    D   it has a large body volume in relation to body surface.

11. Which of these alternatives is **not** a feature of each nephridium of an earthworm?
    A   Closely connected to a rich blood supply
    B   Both osmoregulatory and excretory
    C   Closed by sphincter muscles
    D   Flame cells at the distal end

12. Which of the following activities of the earthworm contributes most to the fertility of the soil? It
    A   secretes calcium carbonate into the food in its gut.
    B   grinds up the food in a muscular gizzard.
    C   burrows deeper into the soil to hibernate.
    D   deposits "worm casts" soil which has passed through its gut.

13. During the copulation of two earthworms the spermatozoids move
    A   into the female reproductive openings of the other worm.
    B   always from the younger to the older worm.
    C   into the clitellum.
    D   to the spermatheca of the other worm.

14. Which of the following is **not** an adaptation for effective gaseous exchange in earthworms?
    A   The body surface is relatively thin
    B   All blood vessels are found deep in the body
    C   The body wall is continuously kept moist
    D   The elongated body supplies an enlarged area surface

15. The blood vessel in the body of *Lumbricus* in which blood flows from back to front is the . . . blood vessel.
    A   intestinal   C   dorsal
    B   subneural   D   ventral

B. Write down the correct term for each of the following statements.

1. The body of the tapeworm except the scolex
2. A single segment of the tapeworm
3. The thick protective layer of the tapeworm
4. The tubes which transport sperms from the testes to the penis
5. The hard protective coat of the hexacanth embryo
6. The dorsal openings of the earthworm which are found from segment 9 backwards
7. The epithelium which lines the coelom of the earthworm
8. The large yellow cells which store fats and glycogen in the earthworm
9. The oxygen carrying pigment which is dissolved in the blood plasma of the earthworm
10. The condition where the sperms are released before the ova are ripe

C. Answer the following questions which relate to the diagram.

1. Of which animal is this a transverse section?
2. What is the name of the part numbered 1?
3. What is the name of the part numbered 2?
4. Mention **one** function of the part numbered 1.
5. Mention **one** nutritional function of the part numbered 2.

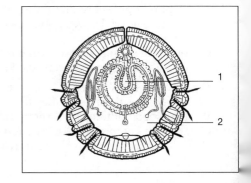

Section B

1. Explain the significance of:
   (a) the flattened form to *Taenia solium* and
   (b) a through-gut and coelom to *Lumbricus terrestris*.

2. An animal has the following features:
   bilateral symmetry
   triploblastic organization
   acoelomate structure
   flame cells
   sucker
   (a) Briefly describe each of these features.
   (b) Name the phylum to which the animal belongs.

3. Answer the following questions on the nutrition of the earthworm.
   (a) Why is the earthworm said to have a through-gut?
   (b) What is the nature of its food and how is the food taken in?
   (c) Name three functions of the calciferous glands.

4. Name the different external openings in the body of the earthworm. Indicate the most important function of each opening as well as the number of the segment on which the opening is located.

5. The following diagram illustrates the blood circulatory system of an animal you have studied.

   Study the diagram and answer the questions which follow.

   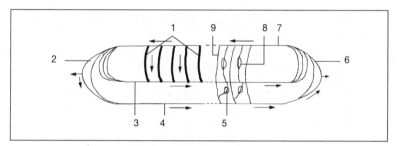

   (a) Is this an open or closed circulatory system?
   (b) Give the name of the amimal in which this particular system is found.
   (c) Identify the parts numbered 1 to 9 on the diagram.
       Which of the numbered parts have valves?
   (d) Mention three differences between an open and a closed blood circulatory system.

6. What is the role played by the digestive tract and the coelom in the nutrition of Annelida?

7. Describe the method of locomotion in the earthworm.

8. Describe the sexual reproduction in the earthworm.

Section C

The tapeworm, *Taenia solium*, leads a solitary existence in the gut of man. Explain how this animal reproduces and finds a new host for its offspring.

HIGHER GRADE
ANSWERS

Section A

A. 1. B   2. B   3. D   4. D   5. D   6. C   7. C
     8. D   9. D   10. C   11. D   12. D   13. D   14. B
     15. C

B.
1. Strobilus
2. Proglottide
3. Tegument
4. Vas deferens
5. Embryophore
6. Dorsal pore
7. Peritoneum
8. Chloragogen cells
9. Erythrocruorin
10. Protandrous

C.
1. Earthworm
2. Typhlosole
3. Coelom
4. To increase the absorption surface of the alimentary canal
5. The alimentary canal is situated there; helps with the transport of food and wastes

Section B

1. (a) **Flattened form**
   The ratio of surface area to volume is large; and the flattened body decreases the distance of diffusion
   A transport system unnecessary; all cells are near food in the gut and also near the external medium
   Gaseous exchange carried out over whole surface area, oxygen and carbon dioxide as well as food diffuse readily from cell to cell
   (b) **Through gut and coelom**
   Newly ingested food is not mixed with partly digested food; and waste matter
   Different parts of the gut have different functions; muscles in wall make peristalsis possible and food moves in one direction
   Undigested residues are egested through the anus

2. (a) *Bilateral symmetry*
   A condition where on half of the organism, from a section taken down its long axis is a mirror image of the other half
   Triploblastic
   Three layers of cells in embryonic stage: the ectoderm, mesoderm and endoderm
   Acoelomate
   Are aminals without a coelom in its mesoderm
   Flame cells
   Cup-shaped cells that occur in animals which by action of cilia; draw fluid waste products into a cavity and then to the exterior
   Sucker
   An organ of attachment in Platyhelminthes
   (b) Platyhelminthes

3. (a) Mouth is the anterior opening for the intake of food; anus is the posterior opening through which undigested food leaves the body
   (b) Soil rich in organic matter; it is first covered with mucus; then drawn into the buccal cavity by suction of pharynx
   (c) It gets rid of excess calcium salts; neutralise organic acids; carbon dioxide is excreted in this way

4. Mouth; on first segment; is for intake of food
   Nephridiopores; on each segment except first three and last one; for the excretion of liquid wastes
   Openings of the spermathecae; between segments 9 and 10, and 10 and 11; store sperms of another worm
   Openings of oviducts; on segment 14; release ova
   Openings of vas deferens; on segment 15; release spermatozoids
   Anus; on the last segment; for the excretion of solid wastes
   Dorsal pores; situated on all segments except first 8 and the last one; secrete coelomic fluid

5. (a) Closed circulatory system
   (b) Earthworm (*Lumbricus*)
   (c) 1. Pseudohearts; valves   6. Posterior capillaries
       2. Anterior capillaries   7. Dorsal blood vessel; valves
       3. Ventral blood vessel   8. Intestinal capillaries
       4. Subneural blood vessel; valves 9. Parietal blood vessels
       5. Nephridial capillaries
   (d) **Open**   **Closed**
       Blood leaves blood vessels   Blood never leaves blood vessels
       Tube-like heart with ostia   No openings in heart
       Does not transport oxygen   Transports oxygen
       Body cavity a haemocoel   Body cavity a coelom

6. The earthworm possesses a through-gut and food moves uninterrupted through it
   Rotting plant material and soil are sucked into the mouth by the muscular pharynx
   The digestive tract is divided into different parts and digestion is more effective as a result of specialisation
   The oesophagus leads to a crop where food is stored and in muscular gizzard it is ground into a pulp
   The food now moves into the intestine; where digestion by enzymes and absorption take place
   Invagination of the typhlosole increase the absorption surface area
   Water is reabsorbed; and undigested wastes are excreted through the anus
   The walls of alimentary canal possess epithelium cells; secreting enzymes
   Longitudinal and circular muscles create peristalsis which drives food forward; and digestion is extracellular
   Coelom filled with coelomic fluid; with useful salts and water
   The nephridia float in fluid which is taken in through the nephrostome. In coiled tube the useful substances are reabsorbed and the coelomic fluid serves as hydrostatic skeleton; supporting peristalsis and facilitates movement of food, thus promotes ingestion

7. The chaetae are used to anchor the worm; and the longitudinal and circular muscles are used for movement; they work against hydrostatic skeleton
   The posterior part of body is anchored by chaetae; and the anterior chaetae are withdrawn; the circular muscles now contract; and the longitudinal muscles relax
   The body gets longer thinner; and the front part of body is pushed forward
   The anterior part of body is now anchored by chaetae; and the chaetae of the posterior part are withdrawn; the longitudinal muscles now contract; and the circular muscles relax
   The body gets shorter and thicker; and the posterior part is drawn forward
   This movement is repeated as a wave running over the body

8. Mating takes place in summer or in spring
   Two worms lie with their ventral sides close together, in opposite directions
   Segments 9 to 11 are opposite the clitellum of the other worm
   A mucus sheath is formed around them
   Sperm from segment 15; move along the sperm groove towards the clitellum; and forced into spermatheca of the other worm

The same occurs in other worm, i.e. sperm are exchanged
The worms now separate
A cocoon is later formed and the clitellum deposits food into it
The worm reverses out of the cocoon; and when it moves over segment 14; ova are deposited in it
When it moves over segment 9 and 10, and 10 and 11, sperms of the other worm are also deposited in it
External fertilisation occurs inside the cocoon
There is a direct development; i.e. no metamorphosis
One or two worms hatch in the cocoon from where they are released; the others are used as food

Section C

In a mature proglottide, sperms are produced in testes
The sperm is passed along genital ducts (vasa efferentia and vas deferens) to the penis; and genital opening which pass sperms on to the vagina of the same proglottide or one positioned nearby
Self fertilisation occurs
From the vagina sperms move along an oviduct; where they fertilise the eggs which are produced in the ovaries
Fertilisation thus internal
The fertilised eggs are supplied with yolk from the yolk gland; and are surrounded with a chitinous shell
A proglottide become filled with eggs; and drops of the strobilus; and is passed out of the body with faeces of the host
The proglottide have a wriggling movement
It disintegrate later; and so the eggs are released
Shortly after fertilisation a zygote starts to develop; and from it a larvae with six hooks; the hexacanth inside the egg
A hard shell, the embryophore develops; and the complete structure is called an onchosphere
Onchospheres remain active for a long time; and they only develop further in an intermediate host; the pig
Onchospheres are taken in when the pig routs in the soil
In the stomach of the pig the hard shell is dissolved; and the hexacanth is released. It bores with hooks through intestinal wall; into a blood vessel and transported with blood to all parts of the body
It develops further in muscle tissue of the pig; into a cysticercus (bladderworm); with proscolex
The pig is said to have measles
If such contaminated, uncooked pork is eaten; the proscolex everts; and attaches itself to the intestinal wall of the host; and forms new proglottides
The life cycle now starts all over

6 Invertebrates – Phylum: Arthropoda

Class Insecta

Example: *Locusta pardalina* – the locust

1. **Habitat**
   Locusts live in regions with **sandy soil**, such as dry grass lands and semi-deserts. From there they **migrate** to adjacent areas.

2. **External structure**
   The locust is **bilaterally symmetrical**, **triploblastic** (three germinal layers) and covered with an **exoskeleton** (skeleton on the outside) of **chitin** protecting it against mechanical injury and desiccation. The skeleton is **strong** and serves for the attachment of muscles making locomotion possible. **Moulting** (ecdysis) is necessary and important for growth.

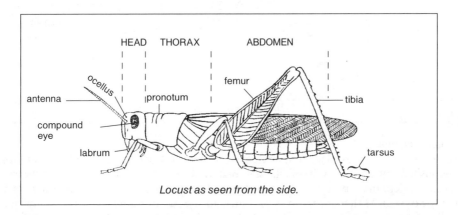

Locust as seen from the side.

   The body is divided into a **head**, **thorax** and **abdomen**. On the six fused segments of the head are **one pair of compound eyes**, **three ocelli** (simple eyes), **one pair of antennae** and the mouthparts, i.e. **one pair of mandibles**, one pair of **maxillae, labium** and **labrum**.
   The **thorax** consists of three segments and on each a pair of **jointed legs** is found. On the **meso-** and **metathorax** are a pair of **spiracles** and a **pair of wings**. The first pair, the **elytra** are leathery, and the wings on the metathorax are soft.
   The **abdomen** consists of **eleven** segments of which the last three are modified for **mating** and **egg laying**. The **dorsal tergum** is joined by the **pleuron** to the ventral **sternum**. On the first abdominal segment there is a **tympanic membrane** on either side. The first eight abdominal segments each have a pair of **spiracles**, leading to the **tracheal system**.

3. **Locomotion**
   The locust moves by **walking** with three pairs of jointed legs. It **flies** with two pairs of wings.

109

4. **Nutrition**
   Locusts are **herbivores**. Sensory organs detect food. **Mandibles** chew off the food with a scissors-like action, while the **maxillae** hold the food. The **maxillary** and **labial palps** taste the food. The **labrum** and **labium** prevent the food from falling out of the mouth. **Ingestion** takes place through the mouth. In the through-gut the food is digested further. There are three main parts:

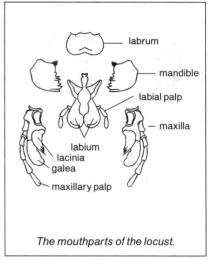

The mouthparts of the locust.

(a) The **fore-gut** consisting of a **buccal cavity** secreting enzymes followed by the **pharynx** and **oesophagus**, joining it to the **crop** in which food is stored and the **gizzard** which grinds the food further.

(b) The **mid-gut** possesses the **gastic caecae** secreting enzymes for extra-cellular digestion. Digested food is absorbed into the blood in the haemocoel for further distribution.

(c) The **hind-gut** is divided into an **intestine**, **rectum** and **anus** through which undigested food leaves the body by **egestion**. Water is mainly absorbed from here.

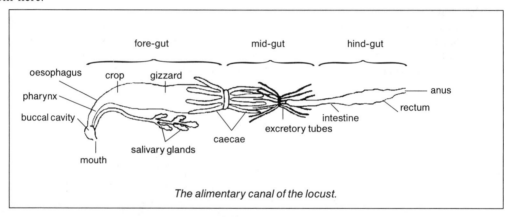

The alimentary canal of the locust.

5. **Gaseous exchange**
   Gaseous exchange takes place in an extensive system of **tracheal tubes**. The circulation of air is accomplished by contraction of the **tergo-sternal muscles** causing a pumping mechanism. Trachea are kept open by **spiral chitinous bands**. In the tips of the tracheoles is a watery liquid in which the oxygen in the air dissolves to diffuse to the surrounding tissues. The internal location of the gaseous exchange surface avoids loss of water and creates a large surface area for gaseous exchange.

6. **Internal transport** (HG only)
   The **open blood system** consists of a **heart** with **ostia** (openings) in the **pericardial sinus**. From there the open ended **aorta** runs, taking blood to the **haemovisceral sinus** of the **haemocoel**. From there blood flows through openings in the **pericardial membrane** back to the heart in the **pericardial sinus**. In this

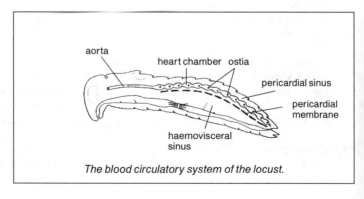

The blood circulatory system of the locust.

way a blood circulation is accomplished. There is **no oxygen carrying pigment** present in the blood and it plays almost no part in transporting respiratory gases. The function of the blood is mainly to transport nutrients and hormones to the surrounding cells and to remove wastes from it.

7. **Reproduction**

Sexual reproduction occurs in the locust and the sexes are separate. Copulation takes place and fertilisation is **internal** when **spermatophores** are placed in the **vagina** of the female. After

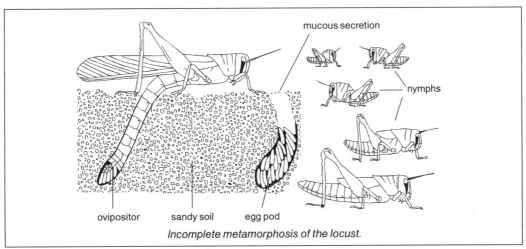

Incomplete metamorphosis of the locust.

fertilisation the egg is supplied with **yolk** and a **shell**. The female makes a hole in the soil with her **ovipositor** and the egg case or pod is covered with a frothy fluid. Locusts are **egg-lying** or **oviparous**. The wingless **nymph** or **hopper** makes its way out of the egg and soil. It resembles the adult locust, but is wingless with a large head. They **moult six times**. **Metamorphosis** is **incomplete** because there are only three stages, i.e. egg, nymph and adult. Complete metamorphosis has four stages, i.e. egg, larva, pupa and adult.

STANDARD GRADE
QUESTIONS

Section A

A. *Various possibilities are suggested as answers to the following questions. Indicate the correct answer.*

1. Arthropoda is
   - A   bilateral symmetrical.
   - B   radial symmetrical.
   - C   central symmetrical.
   - D   asymmetrical.

2. The mouthparts of the locust are
   - A   chewing.
   - B   biting.
   - C   sucking.
   - D   sticking.

3. The shape of the abdomen of the locust changes during inhalation as a result of the
   - A   elasticity of the segments.
   - B   increase in pressure.
   - C   up and down movements of the wings.
   - D   contraction of muscles.

4. How many pairs of spiracles are present on the thorax of the locust?
   A   1
   B   2
   C   4
   D   8

5. Respiratory gases are distributed through the body of the locust by
   A   blood.
   B   skin.
   C   tracheoli.
   D   tissue fluid.

6. The locust breathes by means of
   A   gills.
   B   lungs.
   C   its skin.
   D   trachea and tracheoles.

7. An insect moults
   A   because the body covering has become too old and rigid.
   B   to form wings.
   C   to promote faster movement.
   D   to permit growth.

8. Which of the following animals has a body cavity called a haemocoel?
   A   Locust
   B   *Hydra*
   C   Tapeworm
   D   Earthworm

9. The dorsal part of the insects abdomen is the
   A   pleuron.
   B   sternum.
   C   tergum.
   D   telson.

10. Which of these features is common to all insects and yet found only in insects?
    A   Chitinous exoskeleton
    B   Paired antennae
    C   Simple eyes
    D   Three pairs of legs

11. Which of the following animals possesses an exoskeleton?
    A   *Amoeba*
    B   *Hydra*
    C   Tapeworm
    D   Locust

12. The graph illustrates the growth of

    A   a bird.
    B   a reptile.
    C   a tadpole.
    D   an insect.

13. Which of the following is **not** involved in gaseous exchange in the locust?
    A   Trachea
    B   Muscles
    C   Typhlosole
    D   Tracheoles

14. Hermaphroditism is characteristic of
    A   Protozoa
    B   Coelenterata
    C   Arthropoda
    D   Insecta

15. The external sexual differences between male and female locusts can usually most clearly been seen from the
    A   front parts of the wings.
    B   last three abdominal segments.
    C   middle segment of the thorax.
    D   front part of the head.

16. In which of the following situations is a desert locust most likely to lay her eggs? In
    A   the water of an oasis.
    B   between stones.
    C   gravel at the edge of a stream.
    D   sandy soil.

17. Which one of the following represents the stages of complete metamorphosis in the correct order?
    A   Adult – eggs – pupa – caterpillar
    B   Adult – eggs – caterpillar – pupa
    C   Eggs – adult – pupa – caterpillar
    D   Pupa – eggs – adult – caterpillar

18. The metamorphosis of the locust is
    A   complete, the larvae resembles the adult but is smaller.
    B   incomplete, the larvae differs from the adult.
    C   imcomplete, the larvae develops directly into an adult.
    D   complete, the larvae develops into a pupa.

19. The locust lays eggs and are described as
    A   oviparous.
    B   viviparous.
    C   ovoviviparous.
    D   asexual.

B. Write down the correct term for each of the following statements.

1. The structures which keep the trachea open
2. The type of skeleton of the locust
3. Covers the other mouthparts of the locust
4. Respiratory openings on the abdomen of Insecta
5. The basic material of which the exoskeleton is composed
6. Muscles which work opposite one another
7. The muscles responsible for respiratory movements in Insecta
8. The structure of the female locust used for laying eggs
9. The process of change in the life cycle of an animal
10. The immature stage in the life cycle of the locust
11. The external openings of the tracheal system of the locust
12. Animals with mesodermal cavities in which the internal organs are found
13. The part that covers the first thoracic segment of the locust

C. Match the description in column B which best suits the term in column A.

N.B. A name from column A may be used more than once.

Column A	Column B
Hydra	1. Excretory organs nephridia
Locust	2. Hydrostatic skeleton; segmented
Amoeba	3. Body radially symmetrical
Tapeworm	4. Posseses one pair of antennae
Earthworm	5. Hydrostatic skeleton; tentacles
	6. Head, thorax, abdomen
	7. Movement by cytoplasmic outgrowths
	8. Parasitic
	9. Ecto- and endoplasm
	10. Flame cells for excretion

D. Where does reproduction take place in the following animals? Is it in- or external.

(*a*) *Hydra*
(*b*) Earthworm
(*c*) Locust
(*d*) Tapeworm

E. **Complete the following table:**

	Platyhelminthes	Annelida	Insecta
Example studied	1.	2.	3.
Surface used for gaseous exchange	4.	5.	6.
Are breathing movements used?	7.	No	8.
Way in which gaseous exchange surface area is increased.	9.	10.	11.

F. *Study the following diagrams and answer the questions.*

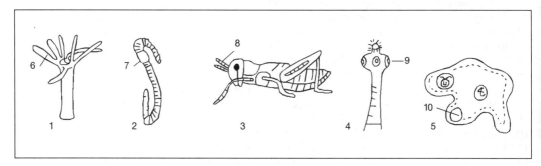

(a) Name the phylum to which the animals numbered 1 to 5 belong.
(b) Give one function of the numbered parts 6 to 10.

G. *The following questions refer to the accompanying organisms. From this group of animals select (write down the letter of the animal only):*

(a) an animal in which blood plays a role in the transport of oxygen.
(b) three which are hermaphroditic.
(c) one where fertilisation is external.
(d) one which moves by pseudopodia.
(e) one which is a stage in a life cycle.
(f) one which is a parasite in man.
(g) one which possesses an exoskeleton.
(h) two which are found in water.

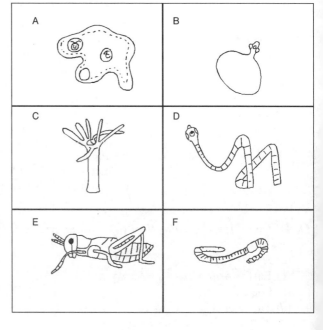

H. *Name an example of an organism in which, or on which, each of the following structures are found and state one function of each.*

   (a) Typhlosole
   (b) Clitellum (saddle)
   (c) Compound eye
   (d) Scolex
   (e) Contractile vacuole

Section B

1. (a) Mention the name of the insect you have studied.
   (b) What type of food does this insect eat?
   (c) Describe the role of its mouthparts in ingesting food.

2. Study the diagram of the head of the locust and answer the questions that follow:
   (a) Identify the parts numbered 1 to 9.
   (b) Which part is responsible for chewing the food?
   (c) What are the part numbered 1 used for?
   (d) What is/are the function(s) of the parts numbered 6 and 7?
   (e) Mention the functions of the parts numbered 2 and 3 respectively.

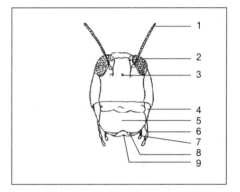

3. Study the diagram which represents the tracheal system of an insect.
   (a) Identify the structures numbered 1 and 2. What are their functions?
   (b) Identify the tubes numbered 3 and 5.
   (c) Identify the structure numbered 4. What is its function?
   (d) Describe how inspiration occurs in an insect.

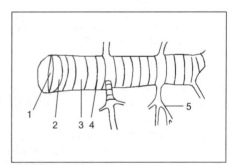

4. Study the diagram of the abdomen of the locust and answer the questions that follow.

   (a) Supply the labels 1 to 9.
   (b) Is this an abdomen of a male or female locust?
   (c) What is the function of the part numbered 7?

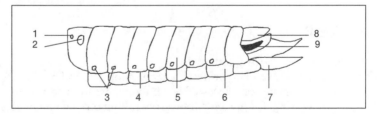

5. Answer the following questions on the gaseous exchange organs in Arthropoda.
   (a) Name the gaseous exchange organs in Insecta.
   (b) How are the organs mentioned in (a), protected?
   (c) Describe how carbon dioxide-laden air leaves the body of an insect.

115

STANDARD GRADE
ANSWERS

Section A

A. 1. A   2. B   3. D   4. B   5. C   6. D   7. D
      8. A   9. C   10. D   11. D   12. D   13. C   14. B
      15. B   16. D   17. B   18. C   19. A

B. 1. Chitin spirals   2. Exoskeleton   3. Labrum
      4. Spiracles   5. Chitin   6. Antagonistic muscles
      7. Tergo-sternal muscles   8. Ovipositor   9. Incomplete metamorphosis
      10. Nymph   11. Spiracles   12. Coelomate
      13. Pronotum

C. 1. Earthworm   2. Earthworm   3. *Hydra*
      4. Locust   5. *Hydra*   6. Locust
      7. *Amoeba*   8. Tapeworm   9. *Amoeba*
      10. Tapeworm

D. (*a*) External   (*b*) External   (*c*) Internal   (*d*) Internal

E. 1. *Taenia solium*   2. Earthworm   3. Locust
      4. Body wall   5. Skin/epidermis   6. Trachea/tracheoli
      7. No   8. Yes   9. Body flattened
      10. Segmental body increases surface area   11. Trachea are branched into finer tracheoli

F. (*a*) 1. Coelenterata   2. Annelida   3. Arthropoda   4. Platyhelminthes   5. Protozoa
      (*b*) 6. Catch food   7. Forms cocoon   8. Sensory organ   9. Attach worm to intestinal wall
      10. Osmoregulation

G. (*a*) F   (*b*) B, C, D, and F   (*c*) F   (*d*) A
      (*e*) B   (*f*) D   (*g*) E   (*h*) A and C

H. (*a*) Earthworm; increase absorption surface area   (*b*) Earthworm; forms cocoon
      (*c*) Locust; movement of far off objects detected
      (*d*) Tapeworm; attachment to intestinal wall   (*e*) *Amoeba*; osmoregulation

Section B

1. (*a*) Locust   (*b*) Plants
   (*c*) *Labrum*: covers and protects other mouthparts; and prevents food from falling out of the mouth
      Mandible: cut off pieces of food; and chew food to form fine particles
      Maxillae: galea prevents food from falling out; lacinia guides the food and pushes it between the mandibles; and the maxillary palps help in selection of food
      Labium: food rests on it; the labial palps assist in the selection of food

2. (*a*) 1. Antenna   2. Compound eye   3. Ocellus   4. Mandible
      5. Labrum   6. Maxillary palp   7. Labial palp
      8. Maxilla   9. Labium

(b) Mandibles
   (c) Sensory organ
   (d) Serve as taste organ and helps with the selection of food
   (e) 2 – detect movement; 3 – distinguish between light and dark and detect nearby objects

3. (a) 1 – Spiracle: opening which leads to trachea; through it gases enter and leave the body
       2 – Chitinous valve; opens spiracle during inspiration and closes spiracle during expiration; which produces a definite circulation of air through tracheal system
   (b) 3 – Trachea   5 – Tracheole   (c) Chitinous ring; to strengthen trachea and keep it open
   (d) *Inspiration*
       The tergo-sternal muscles relax; and the volume of the abdomen increases; and the pressure on trachea decreases
       Air now flows into body; through first 4 pairs of spiracles; whilst the valves of last 6 pairs of spiracles close
       By contraction of the tergo-sternal muscles; pressure on the tracheoli is increased; and the rate of diffusion between tracheoli and tissue cells increases

4. (a) 1. Spiracle   2. Tympanum   3. Spiracles
       4. Pleuron   5. Tergum   6. Sternum
       7. Ovipositor   8. Anal plate   9. Cercus
   (b) Female locust   (c) Make hole in soil for laying eggs

5. (a) Tracheae/tracheoli
   (b) They are sunken into the body; are chitin-lined and spirally strengthened to prevent them from collapsing
   (c) *Expiration*:
       The tergo-sternal muscles contract and the abdomen becomes flatter and thinner
       The pressure on tracheae increases and the valves of first 4 pairs of spiracles are closed; carbon dioxide laden air; is thus forced out of last 6 pairs of spiracles

HIGHER GRADE
QUESTIONS

Section A

A. *Various possibilities are suggested as answers to the following questions. Indicate the correct answer.*

1. Which statement is correct with regard to animals in which cephalisation occurs?
   A   They live parasitically because of a poorly developed nervous system
   B   Slow locomotion in the animals as a result of poor co-ordination
   C   The animals have a concentration of nervous tissue in the head
   D   Nervous tissue is mainly concentrated in the abdomen

2. Which of the following is incorrect? The exoskeleton of Arthropoda
   A   limits the size of the animal.
   B   serves as levers between moveable parts.
   C   prevents the loss of water.
   D   serves as attachment for the endoskeleton.

3. Which one of the following is the most general characteristic of Arthropoda?
   A   Cephalothorax   C   Book lungs
   B   Spiracles   D   Jointed legs

4. Insects are
   A   acoelomatic.   C   diploblastic.
   B   triploblastic.   D   asymmetrical.

5. The gaseous exchange surface of insect is placed internally to
   A   bring it near the gills.
   B   increase air circulation.
   C   be near the air sacs.
   D   prevent water loss.

6. When in flight, the locust
   A   moves both pairs of wings up and down simultaneously.
   B   usually moves both pairs of wings up and down independently.
   C   move only the front wings up and down.
   D   move only the hind wings up and down.

7. Which of the following structures is **not** associated with the collection of food in animals?
   A   Pseudopodia
   B   Peristomium
   C   Mandibles
   D   Tentacles

8. Which of the following groups of structures are concerned with nutrition in animals?
   A   Tentacles, mouth, booklungs, mantle cavity
   B   Labium, spiracles, trachea, hypostome
   C   Maxillae, nematoblast, prostomium, hypostome
   D   Antennae, pseudopodia, contractile vacuole

9. When the locust exhales the
   A   sternum is lifted.
   B   pressure on the abdomen decreases.
   C   valves of the last six pairs of spriracles close.
   D   volume of the abdomen enlarges.

10. Which of the following are concerned with the transport of oxygen-rich air to the body cells of an insect?
    A   Haemocyanin
    B   Spiracles
    C   Erythrocruorin
    D   Tracheoli

11. The significance of incomplete metamorphosis in an insect is that it
    A   allows time for the adult to develop fully.
    B   avoids predators.
    C   ensures a period of dormancy when the larva is inactive.
    D   ensures protection from the cold season.

B. Write down the correct term for each of the following statements.

1. The class of Arthropoda where the body is divided into a head, thorax and abdomen
2. The mesothoracic wings of the locust
3. Animals like the locust which feed on plants
4. The type of alimentary canal which stretches from the mouth to the anus
5. The scientific term for moulting
6. The membrane which divides the haemocoel of the locust into a pericardial and perivisceral sinus
7. Openings with valves in the heart of the locust
8. The packet of sperm cells of the locust
9. Egg   nymph   adult
10. The immature stage in the life cycle of the locust
11. The only cells in the blood of the locust
12. The fluid filled cavity between the gut and the body wall of the earthworm

C. *For each characteristic in column B, select the correct phylum from column A. A phylum from column A may be used more than once.*

Column A
Protozoa
Coelenterata
Platyhelminthes
Annelida
Arthropoda

Column B
1. Proglottide
2. Dorsal pore
3. Tracheal tubes
4. Nematoblast
5. Cysticercus
6. Contractile vacuole
7. Chloragogenous cells
8. Tympanum
9. Amoeboid movement

D. *In each of the following statements one word has been used incorrectly. Spot the incorrect word and write the correct word down.*

1. A coelenteron occurs in all triploblastic animals such as *Hydra*
2. The tapeworm is bilaterally symmetrical, triploblastic, hermaphroditic and saprophytic
3. In *Hydra* spp. the ovaries and testes developed from the glandular cells
4. The tape- and earthworm are diploblastic, i.e. the body develops from an endoderm, mesoderm and ectoderm
5. After oxygen has passed through the trachea of the locust it is transported by the blood to the tissues

Section B

1. Discuss briefly the exoskeleton of Insecta.

2. The questions which follow refer to the diagrams.

    (a) To which phylum does this animal belong?
    (b) To which class does this animal belong?
    (c) Which one of the diagrams (A or B) represents a part of the body of the female animal?
    (d) Identify the parts numbered 1 to 3.
    (e) What is the function of the part numbered 3?
    (f) Is the animal a primary or secondary consumer?
    (g) Name the functions of the parts numbered 4 to 10.

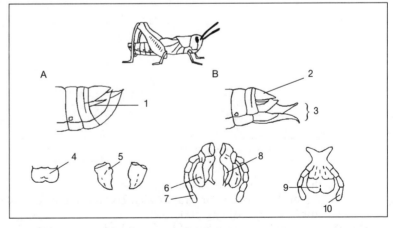

3. State the requirements that an efficient gaseous exchange surface of an animal should comply with.

4. Discuss briefly the three types of body symmetry in animals. Give an example of each type of symmetry.

5. The diagram illustrates the blood system of an animal you have studied.
   (a) To which phylum does this animal belong?
   (b) Is this blood system open or closed?
   (c) Identify the parts numbered 1 to 6.
   (d) Tabulate **two** differences between this type of blood system and another type you have studied.

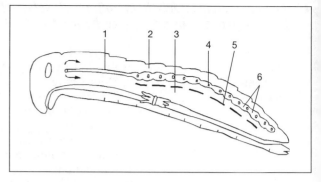

6. Insects possess an open blood circulatory system: the heart pumps blood into the body cavity where it bathes the tissues and organs. Mammals in contrast have a closed blood circulatory system: the heart pumps the blood into a series of vessels throughout the body.
   (a) What are the advantages and disadvantages of an open blood circulatory system?
   (b) The mammal's blood is red and that of insects colourless. Explain the reason for this difference.
   (c) The placenta of mammals contains sinuses. Suggest a reason for this phenomenon.

7. Tabulate **six** differences between the reproduction of Annelida and Insecta.

8. Locusts prefer to lay eggs in soil that are sandy and are slightly moist. Locusts in a cage were given sandy soils with different amounts of water into which to lay their eggs. The female uses her abdomen to make a hole in the soil in which to lay her eggs. A sticky substance is also secreted by the female. This hardens to form an egg pod. The results of the egg laying of the locusts in different soils were recorded as below.

ml water added to 90 g of sandy soil	*Number of egg pods laid*
0	0
2	1
4	2
6	12
8	19
10	20
12	24
14	18
16 (waterlogged)	0

Answer the following questions:
(a) Represent the information graphically. Use the x-axis for "Water added to the soil" and y-axis for "Number of egg pods laid".
(b) What would be the number of egg pods laid if there were 11 ml of water present in the soil?
(c) How much water must be present in the soil for the maximum number of egg being laid?
(d) How much egg pods are laid in waterlogged soil?
(e) Which essential gas will not be present in waterlogged soil?
(f) Which two chemical substances are required for the eggs to develop?
(g) What is the name of the structure used by the female locust to make a hole to lay her eggs?
(h) Why are the eggs laid in the soil rather than on the top of it?

Section C

Compare gaseous exchange of Insecta and Annelida with reference to:
(a) the requirements for an efficient gaseous exchange surface,
(b) the adaptations of the gaseous exchange apparatus for the animal's way of life, and
(c) the mechanism of breathing and gaseous exchange.

HIGHER GRADE
ANSWERS

Section A

A. 1. C   2. D   3. D   4. B   5. D   6. D   7. B
    8. C   9. A   10. D   11. A

B. 1. Insecta   2. Elitrum   3. Herbivore
    4. Through-gut   5. Ecdysis   6. Pericardial sinus membrane
    7. Ostia   8. Spermatophore   9. Incomplete metamorphosis
    10. Nymph   11. Haemocytes   12. Haemocoel

C. 1. Platyhelminthes   2. Annelida   3. Arthropoda
    4. Coelenterata   5. Platyhelminthes   6. Protozoa
    7. Annelida   8. Arthropoda   9. Protozoa

D. 1. Diploblastic   2. Parasitic   3. Interstitial cells   4. Triploblastic   5. Tracheoles

Section B

1. Chitinous exoskeleton; protect against mechanical injury and desiccation
   It has disadvantages for the animal: movement and growth is almost impossible
   The problem is solved by thin membranes; which connects the firm plates and the muscles are attached to the inside of the exoskeleton; they work antagonistically and they result in movement; together with the lever action of the limbs
   The exoskeleton is hard and prevents growth; thus arthropods moult regularly; and before the new skeleton hardens; it grows quickly to a new size

2. (a) Arthropoda   (b) Insecta   (c) B   (d) 1 – cercus   2 – anal plate   3 – ovipositor
   (e) Make hole in soil in which eggs are laid
   (f) Primary consumer; it is plant-eating/herbivore
   (g)   4 – protect the mouthparts behind it   5 – cuts off the food
         6 – holds food   7 – touch organ
         8 – holds food   9 – prevents food from falling out of the mouth
         10 – organ of touch

121

3. The gaseous exchange surface must be large and moist
It should have an effective transport system for gases
It should be thin-walled and in contact with respiratory gases

4. *Asymmetry*: no definite form; and cannot be divided into two identical halves, e.g. *Amoeba*
Radial symmetry: organs radially arranged; there are no left or right hand sides; and can be cut along more than one plane; to produce two identical halves, e.g. *Hydra*
Bilateral symmetry: has a definite front, back, left, right, top and bottom side; and can be cut along one plane; to produce two identical halves; e.g. all vertebrates

5. (*a*) Arthropoda (*b*) Open blood system
   (*c*) 1 – aorta   2 – pericardial sinus
   3 – haemocoel visceral sinus   4 – ostium
   5 – pericardial membrane   6 – heart chambers
   (*d*) **Open blood system**   **Closed blood system**
   Blood bathes organs   Blood travels in blood vessels to the organs
   Blood plays no part in the   Blood transports oxygen and carbon
   transport of respiratory gases   dioxide to and from the organs

6. (*a*) *Advantages*:
   Easier exchange of gases, nutrients and excretory products; when organs are bathed in blood, i.e. no intervening membranes
   Disadvantages
   Special breathing system is required; to bring respiratory gases to the deeper lying tissues
   (*b*) Insect has no pigment; for transporting respiratory gases. It has a system of tracheal tubes which supplies respiratory gasses, directly to the tissue cells
   Mammals have lungs; blood has to transport respiratory gases; and therefore the pigment haemoglobin is required; which is red in colour due to haemoglobin
   (*c*) Quick exchange of respiratory gases and nutritional elements as well as wastes take place

7. **Insecta**   **Annelida**
   Metamorphosis   Direct development
   Internal fertilisation   External fertilisation
   Unisexual   Hermaphrodite
   Eggs protected in egg pod   Eggs protected in cocoon
   Many eggs laid at a time   Few eggs in cocoon at a time
   Sometimes parental care, e.g. honey bee   No parental care

8. (*a*)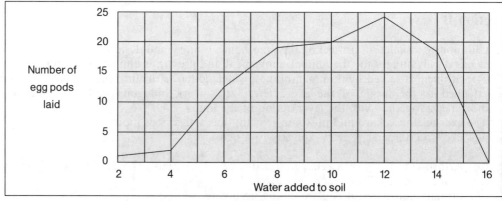

   (*b*) 22 egg pods   (*c*) 12 ml of water
   (*d*) No eggs are laid in waterlogged soil   (*e*) Oxygen
   (*f*) Oxygen and water   (*g*) Ovipositor
   (*h*) For protection during unfavourable conditions

Section C

(*a*) *Requirements for efficient area for gaseous exchange*
It must have a large surface area; it must be moist and thin-walled
An effective transport system of oxygen; and carbon dioxide must be present; as well as an effective ventilation mechanism

Insecta

(*b*) *Adaptation of apparatus for gaseous exchange*
To prevent desiccation in terrestrial environment; the gaseous exchange surface is sunken into the body
The apparatus consists of large number of tracheal tubes; which ramify throughout entire body and come in close, direct association with body cells
The air sacs; and tracheoli possess large gaseous exchange surface; and blood plays a minor role
Oxygen diffuses through walls of tracheoli to cells; and carbon dioxide is removed in same way
The trachea are kept open and strengthened by chitin bands; and a lining of tracheoli is thin and permanent; and the air reaches air tubes via the spiracles

(*c*) *Mechanism of breathing and gaseous exchange*
During **inspiration** the tergo-sternal muscles relax; and the volume of the abdomen increases; and the pressure on trachea decreases
Oxygen-laden air flows in through first 4 pairs of spiracles; while the valves of last 6 pairs of spiracles close
By muscular contraction of the trachea; pressure on tracheoli increases; and the rate of diffusion between tracheoli; and tissue cells is increased
During **expiration** the tergo-sternal muscles contract; the abdomen becomes flatter and thinner; and the pressure increases
The valves of first 4 pairs of spiracles are closed; and carbon dioxide-laden air forced out of last 6 pairs of spiracles

Annelida

(*b*) *Adaptation of apparatus for gaseous exchange*
No special organs for gaseous exchange
It takes place through skin; which is moist and well supplied with blood capillaries
It is kept moist by mucus; from goblet cells; and from dorsal pore
The large body exposes a large surface area; the epidermis is one layer thick and it has a thin cuticle
The blood transports gases; with the pigment erythrocruorin
The extensive blood system; supplies the cells with oxygen
It has no ventilation system; the movement from place to place supplies a new environment

(*c*) *Mechanism of breathing and gaseous exchange*
It has no special mechanism
Oxygen from the air dissolves in the moisture of the skin; and diffuses into the blood capillaries; and taken to cells
Carbon dioxide dissolves in blood plasma; and is transported to the skin; and diffuses from blood through skin to the outside

7 Vertebrata – Phylum Chordata Subphylum Vertebrata

A. Class Osteichthyes (Bony fishes), e.g. *Tilapia* sp.

1. Habitat
The bony fish is found in **water** and is **streamlined**.

2. External chracteristics
The body is divided into **head**, **trunk** and **tail**. On the head are a **mouth**, two large **eyes**, two external **nostrils** for smelling and an **operculum** (a bony plate which covers the gill-slits) on each side of the head.

The **trunk** is bilaterally flattened and covered with **dermal scales**. The **paired** fins are the **pectoral** and **pelvic** fins. The **unpaired** fins are the **ventral** (anal and **caudal** fins. The **dorsal** fin can be paired or unpaired. Ventral, just before the anal fin opens the **cloaca**. On either side of the body is the **lateral line** (a group of sensory organs) which detects **streaming movements** and **vibrations** in the water with which the fish determines its **depth**. The **tail** is muscular.

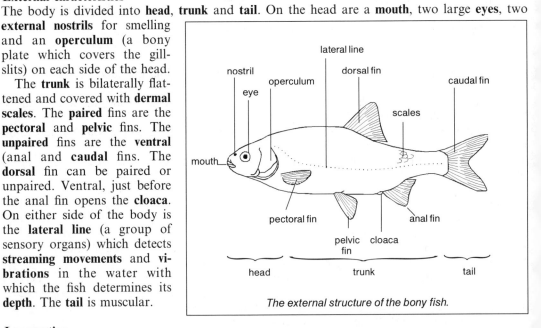

The external structure of the bony fish.

3. Locomotion
The **fins** control **locomotion** which is caused by the muscular **caudal** (tail) **fin** of the fish. The **dorsal** and **pelvic** fins serve to keep the body upright when the fish makes a sudden change of direction. The **pectoral** and **pelvic** fins **balance** and **steer** the fish as it moves. The caudal fin curves **slowly outwards** and then **rapidly** back to the midline, the tail thus presses backwards against the water and in this way the fish is driven forward. This action is repeated on the other side causing a further forward movement of the fish. The **swim bladder**, a bag filled with air, which is found in the body of the fish and is an outgrowth of the **pharynx**, is responsible for **depth control**. Changes in **air volume** controls the buoyancy of the fish.

4. Nutrition
Some fishes are **herbivorous** (plant-eating), while the vast majority are **carnivorous** (meat-eating). Some are even **omnivorous** (eating both plant and animal matter). Herbivores live on **phytoplankton** (algae) which is held back by the gill rakers as water passes over the gills. **Carnivores** hunt down their prey and catch it. The prey is swallowed whole.

5. Gaseous exchange
Gaseous exchange takes place by obtaining dissolved oxygen from the water by means of **gills**.

Water enters the mouth, flows over the gills and out through the **gill-slits**. The mechanism is as follows: operculum closes, mouth opens, floor of buccal cavity lowers, buccal cavity enlarges, pressure inside decreases and water enters through the mouth. The mouth and gullet close, floor of buccal cavity raises, gill-slits open, buccal cavity becomes smaller, pressure increases, the operculum opens and water flows over the gills and passes out through the gill-slits. Dissolved oxygen **diffuses** into the **blood capillaries** of the gill-filaments and **carbon dioxide diffuses** into the water and is carried away. Oxygen is transported by blood to all parts of the body.

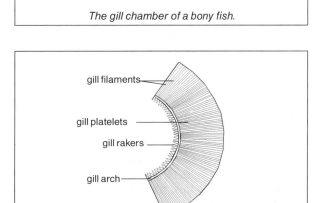

The gill chamber of a bony fish.

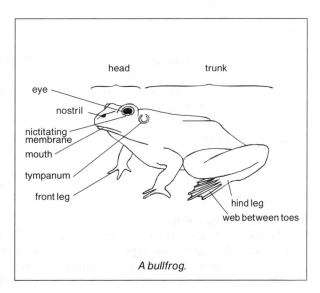

A single gill.

6. Reproduction

Fish are **unisexual** and **sexual reproduction** takes place. **Eggs** are released via the **cloaca** (a chamber at the end of the alimentary canal in which the contents of the canal, the kidneys and sexual organs are released) of the female into the water and the male discharges its **sperms** over the eggs. **Fertilisation** is external. The eggs are supplied with **yolk** and is used during the development of the growing fish until it can fend for itself. Fishes are **oviparous** (egg laying). **Ovoviviparous** (eggs are retained and hatch in mother's body and then give birth to live offspring) types also occur but are not very common (e.g. guppies – *Poecilia reticulata*).

B. Class Amphibia (Frog), e.g. *Rana* sp.

1. The frog lives on **land** and in **water**. **Reproduction** always occurs in water.

2. **External characteristics**

The body is divided into a **head** and **trunk**. There is no neck and tail. The head is broad and dorso-ventrally **flattened**. Two external **nostrils** occur above the wide mouth. Two prominent **protruding eyes** are situated almost on top of the head. Each eye is protected by two movable **eyelids** and a **nictitating membrane**. Behind each eye is a **tympanum** for hearing. The trunk is broad in the middle and tapers towards the **cloaca** at the hind end. Each front limb consists of an **upper arm**, **forearm**, and a **hand** with **four digits**. Each hind limb consists of a **thigh**, **shin** and **foot** with webbed toes. There is a webb between the toes. The bones of the foot and ankle are elongated. The **skin** is **slimy** and is kept moist by many mucus **glands**.

A bullfrog.

3. Locomotion

The adult frog **moves** by **swimming**, **walking** (waddle) and **jumping**. The tadpole can only swim. When jumping the frog folds its hind limbs under its body and then suddenly straightens them. The force pushes the animal forward into the air. The frog lands on its short front limbs which absorb the shock when landing. The frog swims by pushing its **webbed** hind feet against the water.

4. Gaseous exchange

During its development the frog breathes in various ways. In the tadpole **gaseous exchange** occurs by means of **external gills**. At a later stage **internal gills** develop with a **spiracle** to the outside. As the tadpole grows, **lungs** start to develop. The **mucus membrane** of the buccal cavity, the **skin** and **lungs** are used for gaseous exchange. **Ventilation** is achieved by lowering and raising the floor of the mouth. **Blood** transports the respiratory gases. **Mechanism** of lung breathing (inspiration): the **glottis** closes, the mouth is kept closed, the floor of the buccal cavity is lowered and cavity enlarges, pressure in it decreases and air flows into buccal cavity through the nostrils; the glottis opens, the floor of the buccal cavity is raised, the nostrils are shut and air is forced into the lungs, exchange of oxygen and carbon dioxide occurs.

5. Nutrition

Different stages of the frog feeds differently. The young **embryo** survives on the **yolk** of the egg. In early development the **tadpole** scrapes off plant material from water plants. At this stage it is **herbivorous**. At a later stage the diet consists of **insects** and the frog is now **carnivorous**. The tongue is attached at the front of the mouth and has a sticky tip. The tongue is flicked out and the insect, which adheres to it, is placed at the back of the pharynx and **swallowed whole**.

6. Reproduction

Sexes are **separate** and **mating** takes place in water. The eggs are **laid** in shallow water and the male discharges his **sperms** over the eggs. Fertilisation is **external** and frogs are **oviparous**. The egg is protected by a **jelly-like layer** containing albumen. The developing embryo is nourished with **yolk** stored in the egg. A **larva** (tadpole) develops from the fertilised egg and undergoes several changes until it reach the adult stage. **Metamorphosis** thus occurs.

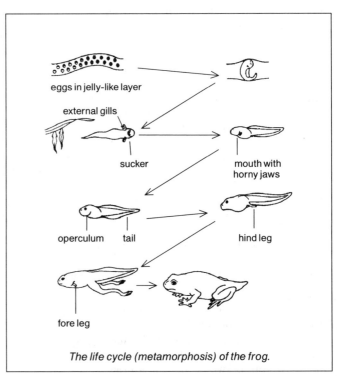

The life cycle (metamorphosis) of the frog.

C. Class Reptilia (Lizard), e.g. *Lacerta* sp.

1. Habitat

Lizards are found in **cracks** in **rocks** or walls, sheltering under leaves, in caves and in the ceilings of most houses.

2. External characterisics

The body is divided into a **head**, **neck**, **trunk** and **tail**. The head is triangular with the **mouth** at the front end. Two **nostrils**, two **eyes**, each with two **eyelids** and a **nictitating membrane** are found on the head. The **tympanic membrane** just behind the eyes is slightly sunken. The neck is short and the trunk is broader than the rest of the body. Two pairs of short **pentadactyle limbs** are attached to the sides of the body. The **cloaca** is found on the ventral side at the hind end of the trunk. Behind the cloacal aperture is a long tail. The body is covered with a dry skin and **horny scales** which overlap like **roof tiles**. The scales on the head are larger, they do no overlap and are known as **shields**.

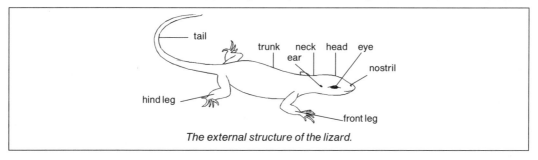

The external structure of the lizard.

3. Locomotion
The lizard can **run, walk** and **climb**. The limbs carry the body off the ground. The limbs work across and alternatively and are helped by the side-to-side movement of the tail.

4. Gaseous exchange
Gaseous exchange takes place in the **lungs**. Air is taken in through the external nostrils. A **diaphragm** is **absent** and the **movements of the ribs** move the air in and out of the lungs. The mechanism is as follows: With **inspiration** the intercostal muscles contract and lift the ribs forward, the body cavity enlarges and the lower pressure let the air flow in. With **expiration** the intercostal muscles relax with the result that the body cavity becomes smaller with an increase in pressure. Air is forced out. In the lungs **oxygen diffuses** into the blood and **carbon dioxide** from the blood into the lungs.

5. Nutrition
Lizards are **carnivorous** and feed on small **insects** and **worms**. The prey is caught by shooting out its long and **sticky tongue**. **Jaws** hold the prey which is swallowed **whole**.

6. Reproduction
During **sexual reproduction** there is courtship behaviour. **Mating** occurs and the **cloaca's** of the male and female are put together and **fertilisation** is **internal**.

Eggs, with a tough **leathery shell**, is laid in loose soil (oviparous). The eggs are **hatched** by the **heat of the sun**. There is **no metamorphosis** and the young resemble the adults. Some types are **ovoviviparous**.

D. Class Aves (Pigeon), e.g. *Columba* sp.

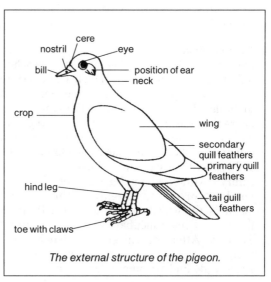

The external structure of the pigeon.

1. Habitat
Pigeons are **veld**, **wood** and **forest** dwellers.

2. External characteristics
The body, covered with **feathers**, is divided into **head, neck, trunk** and **tail**. On the head is a pointed **beak**. Near the base of the beak, just in front of the **cere**, are two nostrils. Both eyes have upper and lower **eyelids** and a **nictitating membrane**. Behind each eye is an **external ear opening** covered at the body surface by small feathers. The **neck** is short and can turn in many directions. The widest part of the body is the **trunk**. It bears **two wings** and **two** hind limbs. Each wing has three main parts; the **upper arm**, **fore arm** and

hand with three **fingers** and is covered with **quill feathers** to increase its surface area. The **hind limbs**, the legs, are used for walking. Each has four digits, three are directed forwards and one to the back. The leg ends in claws and are covered with **scales**. The tail is short and bears **tail quills**.

3. Locomotion

The pigeon moves by **walking** and **flying**. The **wings** are used for flying. The downward stroke of the wing provides the lifting force and forward thrust when it pushes against the air. With the upward stroke the wing is bent at the wrist to cleave the air. The hind legs are so placed as to balance the weight of the body between them while walking. The **wing quills** enlarge the surface area considerably and the **contour feathers** gives shape to the body. In the **epidermis** are **feather follicles**. In each follicle is a **feather papillae** supplied by **blood vessels** and **nerves** responsible for the growth of the feather.

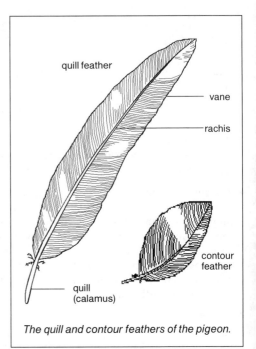

The quill and contour feathers of the pigeon.

4. Gaseous exchange

The pigeon breathes by means of **lungs** and ventilation is accomplished by movements of the rib-cage. The **air sacs** serve as air reservoirs and lowers the specific gravity of the animal. When the animal **walks** the **abdominal** and **intercostal muscles** contract, the sternum is lifted and air is forced out. With the relaxing of the muscles the body cavity enlarges and air moves in. When the animal **flies** the **vertebral column** moves up and down with the contraction and relaxation of the muscles. With **inspiration** air flows into the lungs and air sacs and diffuses into the air sacs in the legs. With **expiration** air is forced out from the lungs and the air from the air sacs fills the lungs. In this way the lungs are always filled with air.

5. Nutrition

The pigeon is a **seed-feader** and the beak is adapted to pick it up from the ground. A special stomach, the **gizzard**, helps with the breaking up of seeds before it is digested further in the stomach.

6. Reproduction

During **sexual reproduction** there is a **courtship**, followed by mating when the **cloacal aperture** of the male is joined to that of the female and the **seminal fluid** is carried to the female and **internal fertilisation** occurs. The two **eggs** are laid in a simple nest. An egg consists of a **calcareous shell**, two **shell membranes**, the embryo with **yolk sac** (which supplies the embryo with food in the form of **albumen**), **amnion** (protection against shock) and **allantois** (in which wastes are stored). Both male and female incubate the eggs for about 21 days. After the chicks have hatched the parents care for them (parental care) until they learn to fly and become independent.

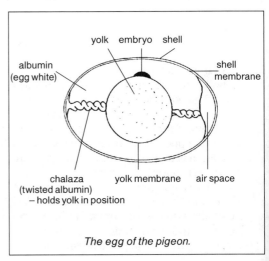

The egg of the pigeon.

E. Class Mammalia (Mammals) – The Rat, e.g. *Rattus* sp.

1. External characteristics

The body is divided into a **head**, **neck**, **trunk** and **tail** and is covered by **hair**. The rat has two pairs of limbs. Each of the **front legs** has four **toes**, each with a claw. Each of the **hind legs** has five toes, each with a **claw**. On the head are two **ears** with **pinnae** directed to the front. At the front of the head below the ears are two eyes, each with a movable upper and lower **eyelid** and a **nictitating membrane**. At the tip of the pointed **snout** are two external **nostrils** just above the split upper lip of the mouth. On each side of the split upper lip are long sensitive **whiskers**. The long **tongue** acts as an organ of taste and helps with eating. Both the upper and lower jaws have **teeth** situated in sockets. The **neck** is short and thick. On the ventral surface of the female's trunk there are six pairs of **teats**. The **genital opening** (vagina), **anus** and **urinary opening** are separate in the **female** and are found between the hind legs. In the male the **urinary** and **sexual opening** open jointly at the **tip** of the **penis**. The **scrotum** holds the **testes** outside the body between the hind legs. The **tail** is long and ends in a thin point.

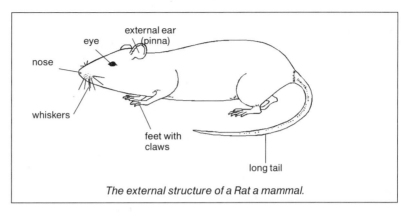

The external structure of a Rat a mammal.

2. Reproduction

During **copulation** the **penis** of the male is inserted into the **vagina** of the female. The **sperms** are discharged and fertilisation occurs **internally** in the oviducts. The **zygote** develops into the **embryo** in the **uterus** of the female. The **placenta**, an intimate physiological connection between the embryo and the mother, supplies food and respiratory gases and removes wastes and carbon dioxide from the embryo. After a **gestation period** the young are born alive. Mammals are **viviparous**. Mammals **suckle** their young for a while on milk produced by **mammary glands** of the mother. Mammals show maximum **parental care** (postnatal care).

STANDARD GRADE
QUESTIONS

Section A

A. Various possibilities are suggested as answers to the following questions. Indicate the correct answer.

1. Internal fertilisation always occurs in animals that are
   - A viviparous.
   - B oviparous.
   - C unisexual.
   - D bisexual

2. Which of the following is a common feature of fishes, birds and reptiles?
   - A Warm-blooded
   - B No metamorphosis
   - C Diaphragm is present
   - D Eggs are protected by a hard shell

3. Which of the following groups of animals undergo metamorphosis during their development?
   - A Osteichthyes
   - B Amphibia
   - C Aves
   - D Mammalia

4. In which animals are ribs and intercostal muscles **not** used during breathing?
   - A Lizards
   - B Pigeons
   - C Rabbits
   - D Frogs

5. Which of the following is common to Osteichthyes, Reptilia and Aves?
   A   Possesses mucus glands and/or oil glands in skin
   B   Eggs have a leathery or calcareous shell
   C   Ovoviviparous
   D   Scales are found on the body or parts of the body

6. Hermaphroditism is characteristic of
   A   Vertebrata   C Arthropoda
   B   Coelenterata   D Insecta

7. Birds and earthworms are both
   A   hermaphroditic.   C   internally fertilised.
   B   oviparous.   D   ovoviviparous.

8. Which animals usually lay fertilised eggs?
   A   Fish and lizards   C   Pigeons and frogs
   B   Lizards and pigeons   D   Frogs and fish

9. A muscular diaphragm is found in
   A   all vertebrates.   C   reptiles.
   B   mammals and birds.   D   mammals.

10. Which characteristic applies to fishes, amphibians and reptiles?
    A   All of them live in or near water   C   All of them are covered by scales
    B   All of them are unisexual   D   No metamorphosis occurs

11. In which of the following animals does the nose exclusively serve as an organ of smell?
    A   Fishes   C   Birds
    B   Mammals   D   Reptiles

12. The graph below indicates the relationship between environmental temperature and the animal's activity. Indicate which two animals could be represented here

	Animal A	Animal B
A	Frog	Hare
B	Lizard	Frog
C	Rabbit	Lizard
D	Rabbit	Pigeon

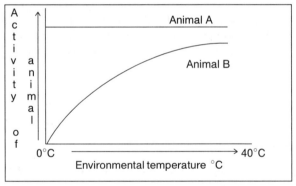

13. Which of the following is the most acceptable explanation of the term "ectothermic"?
    A   Always hibernates during winter
    B   Must live in or near water
    C   Has a constant body temperature
    D   Body temperature varies with variation in temperature of the habitat

14. The animal which has **no** cloaca is the
    A   pigeon.   C   rabbit.
    B   frog.   D   lizard.

15. Scales are found
    A   on all vertebrates.   C   only in fishes and reptiles.
    B   only in fishes.   D   on fishes, reptiles and birds.

16. The Class Osteichthyes is characterised by
    A an absence of an operculum.
    B paired fins present only.
    C an operculum and internal skeleton of bone and cartilage.
    D unpaired fins present only.

Questions 17 and 18 refer to the diagram of a fish.

17. Which number indicates the paired pectoral fins?
    A 3
    B 5
    C 8
    D 10

18. State which number indicates a sensory organ.
    A 1
    B 2
    C 4
    D 9

19. Fishes have
    A paired anal fins.
    B unpaired pectoral fins.
    C paired pelvic fins.
    D paired dorsal fins.

20. On land a frog breathes mainly by means of its
    A lungs.
    B skin.
    C internal gills.
    D mouth.

21. The digestive, reproductive and excretory system of the frog opens in a common cavity known as the
    A kidney.
    B pancreas.
    C cloaca.
    D bladder.

22. The frog belongs to the Class.
    A Amphibia.
    B Chordata.
    C Vertebrata.
    D Reptilia.

23. External fertilisation occurs in
    A reptiles.
    B insects.
    C frogs.
    D birds.

24. Which of the following is true for both frogs and reptiles?
    A Body suspended between the legs
    B No eyelids, only a nictitating membrane
    C Head, neck and trunk clearly visible
    D Pigment cells in the skin of some types

25. The characteristic which classifies this animal as belonging to the class Reptilia, is that it has
    A webbed toes.
    B a naked skin.
    C a skin covered with scales.
    D five fingers and four toes.

26. Birds differ from all other vertebrates because they
    A are warm-blooded.
    B are covered with scales.
    C possess air sacs.
    D possess eyelids.

27. Which one of the following characteristics is **not** representative of birds?
    A Oviparous
    B Ovary
    C Larynx
    D Cloaca

28. Indicate the part of the bird's body where the type of feather represented in the figure, occurs.

    A Along the neck
    B Only near the wing tip
    C Along the wing near the body
    D Along edges of the wings and tail

    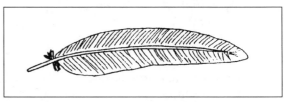

29. A characteristic of the eggs of vertebrates which are fertilised externally, is the
    A small yolk quantity.
    B absence of porous shells.
    C chalaza.
    D presence of an air space.

30. Which of the following is **not** a characteristic of Mammalia?
    A Being warm-blooded
    B Sexes separate
    C Embryo develops inside uterus
    D Oviparous reproduction

31. Which of the following animals get their food by sucking milk?
    A Lizard
    B Tortoise
    C Mouse
    D Tadpole

32. Mammals differ from all other vertebrates in having
    A a constant body temperature.
    B two pairs of limbs.
    C a diaphragm.
    D eyelids.

33. Which characteristic can be found only in mammals?
    A Warm-blooded
    B Skin covering of hair
    C Internal fertilisation
    D External auditory opening

34. Which one of the following features is also characteristic of animals other than mammals?
    A Pinnae
    B Mammary glands
    C Endothermic
    D Hair or fur

B. Write down the correct term for each of the following statements.

1. The swimming and balancing organs of the fish
2. An air-filled sac, an outgrowth from the oesophagus (gullet), inside the body of bony fish
3. An organ sensitive to vibrations and pressure changes in the water surrounding a fish
4. The respiratory organs of the fish
5. The flat semi-circular cover on each side of the body behind the head of the fish
6. The nutritional substance in the egg of the fish on which the embryo exists during early development
7. The scales on the head of a reptile
8. Supplies the bird embryo with food
9. The special sac for storing the embryo's excretory products
10. The fluid filled membrane that surrounds the developing bird's embryo
11. The special structure for nutrition of the embryo of a mammal

C. Complete the following table.

	Osteichthyes	Aves
Example studied	1.	2.
External or internal fertilisation	3.	4.
Protection membrane of the embryo	None	5. & 6.
Food supply to the embryo	7.	8.
Embryo's disposal of nitrogenous wastes	by ... 9. ... in the water	collects in ... 10. ...
Parental care for young (Yes/No)	11.	12.

Section B

1. Describe **five** ways in which fish are adapted to an aquatic life.

2. Study the diagram and answer the questions that follow.
   (a) To which phylum, subphylum and class does the fish, which you have studied, belong?
   (b) Identify the parts numbered 1 to 8 and give one function of each part.

3. Describe the movement of fish through the water with reference to the functions of the various fins.

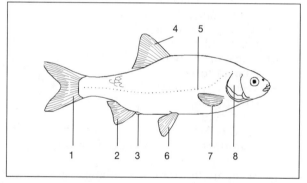

4. Answer the following questions on the gaseous exchange of the fish you have studied.
   (a) Name the gaseous exchange organs of the fish.
   (b) How are the structures, named in (a), protected?
   (c) How is a constant flow of water over these structures, mentioned in (a), maintained?
   (e) How are the respiratory gases transported?

5. Answer the following questions in connection with the reproduction of the fish.
   (a) Is the fish uni- or bisexual?
   (b) In which organs are the
      (i) sperm, and
      (ii) eggs formed?
   (c) Where does fertilisation take place?
   (d) On what does the developing embryo subsist in the egg?
   (e) How and where do the eggs usually incubate?
   (f) Does the small fish undergo metamorphosis? (Yes or no).
   (g) Why are so many eggs released?

6. Draw a labelled diagram to illustrate the external structure of an adult frog.

7. How does the
   (a) tadpole, and
   (b) adult frog move forward?

8. During its lifetime the frog feeds by different methods. How does it takes place during:
   (a) embryonic development,
   (b) the tadpole stage, and
   (c) the adult frog?

9. At different stages in its life the frog breathes in different ways. Where does breathing occur during the following stages?
   (a) The tadpole.
   (b) The adult frog.

10. Answer the following questions on the frog.
    (a) Is the frog uni- or bisexual?
    (b) Is the reproduction of the frog sexual or asexual?
    (c) How and where does fertilisation take place?
    (d) How are the eggs protected and from what does the developing embryo subsist?
    (e) What is the stage, in the life cycle of the frog called which develops from the egg?
    (f) What type of metamorphosis is found here? Explain briefly.

11. Study the diagram and answer the questions which follow.
    (a) To which class does this animal belong?
    (b) What is the habitat of this animal?
    (c) Identify the parts numbered 1 to 6.
    (d) Name the body regions A to D.
    (e) What is the body covering at
        (i) A, and
        (ii) C called?
    (f) How does the animal move forward?

12. (a) What are the lizard's breathing organs called?
    (b) How is air moved into the organs mentioned in (a)?

13. (a) What is the type of nutrition in lizards called?
    (b) How does the lizard capture its prey?

14. The lizard reproduces sexually.
    (a) Are lizards unisexual or bisexual?
    (b) Where and how does fertilisation occur?
    (c) Where and how does the embryo develop?

15. Study the diagram of the pigeon and answer the following questions:
    (a) To which class does this animal belong?
    (b) Identify the parts numbered 1 to 6
    (c) What type of body covering is found on the structure numbered 6?
    (d) What type of feathers are found on the structure numbered 4?
    (e) What are the functions of the parts numbered 4 and 5?

16. Name the adaptations of the pigeon for flight.
17. Make a labelled diagram of a quill feather.
18. (*a*) With what does the pigeon breathe?
    (*b*) Describe the role of the air sacs in breathing of the pigeon.
19. (*a*) Which type of reproduction occurs in the pigeon?
    (*b*) Is the pigeon uni- or bisexual?
    (*c*) Where does fertilisation takes place?
20. The diagram shows the egg of a pigeon about three days after laying
    (*a*) Identify the parts numbered 1 to 9.
    (*b*) Give the functions of the parts numbered 1, 4 and 7.

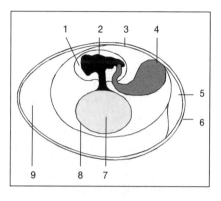

21. Briefly describe the external characteristics of the small mammal you have studied.
22. (*a*) Where does fertilisation in mammals occur?
    (*b*) How are
        (i) the embryo, and
        (ii) the infant mammal fed?

STANDARD GRADE
ANSWERS

Section A

A. 1. A   2. B   3. B   4. D   5. D   6. B   7. B
    8. B   9. D   10. B   11. A   12. C   13. D   14. C
    15. D   16. C   17. A   18. C   19. C   20. A   21. C
    22. A   23. C   24. D   25. D   26. C   27. C   28. D
    29. B   30. D   31. C   32. C   33. B   34. C

B. 1. Fins   2. Swim bladder   3. Lateral line
    4. Gills   5. Operculum   6. Yolk
    7. Shields   8. Yolk sac   9. Allantois
    10. Amnion   11. Placenta

C. 1. Fish   2. Pigeon   3. External
    4. Internal   5. Amnion   6. Chorion
    7. Yolk   8. Yolk and albumin   9. Diffusion
    10. Allantois   11. No   12. Yes

Section B

1. The fish is streamlined; offering least resistance to water
   The skin is covered with scales; impermeable to water; with the free ends pointing backwards offering no resistance to water
   Locomotion is by means of the powerful **caudal** fin
   The **pelvic** and **pectoral** fins are used to steer and balanced the body
   The **dorsal** and **ventral** fins serve to keep the body upright
   Gaseous exchange is by means of gills

The lateral line detects pressure and currents changes in water
The swim bladder is responsible for depth control

2. (*a*) Phylum Chordata; Subphylum Vertebrata; Class Osteichthyes
   (*b*) 1. Caudal fin; locomotion
       2. Anal fin; keeps fish upright
       3. Cloaca; reproduction and excretion
       4. Dorsal fin; keeps fish upright
       5. Lateral line; sensitive to pressure and currents
       6. Pelvic fin; balance of fish
       7. Pectoral fin; balance and stops fish
       8. Operculum; protects gills

3. The caudal fin slowly moves from the midline outwards; and rapidly back towards the midline; it presses against the water; and the fish is pushed forward. The movement is repeated rhythmically. The paired pectoral and pelvic fins; helps to balance; and steer the fish; and the unpaired dorsal and anal fins keep fish upright; when it changes direction

4. (*a*) Gills
   (*b*) The gills are in a gill-chamber for protection; and gill-slits are internal towards the buccal cavity; on the outside the gills are protected by an operculum
   Gill-rakers prevent food particles from damaging the delicate gills
   (*c*) The suction and pumping action of the floor of buccal cavity; which is lowered and raised and the operculum which moves in and out
   (*d*) Haemoglobin; in erythrocytes; transports oxygen; and carbon dioxide dissolves in blood plasma

5. (*a*) Unisexual (*b*) (i) Testes (ii) Ovaries;
   (*c*) External; in the water (*d*) Yolk
   (*e*) By the heat of the sun; in shallow water (*f*) No
   (*g*) Eggs serve as food for other animals; chances thus improved; so that a fair number reach adulthood

6.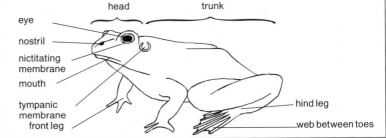

7. (*a*) Swim; by wrigling movements; of body and tail
   (*b*) Swim with webbed hind legs; pushing against the water; front legs used for steering
   Walk on all four legs
   Jump by pushing the hind legs against the ground; the body is propelled forward; front legs absorb the landing shock

8. (*a*) Yolk in egg
   (*b*) Within a week; a tough jaw with horny teeth; scrape off vegetative food from water plants; herbivorous; eater true jaws; with teeth in upper jaw only; catch insects; carnivorous
   (*c*) Tongue long and sticky; flicks out and catches insects; and brings it to the mouth; and swallowed whole

9. (*a*) At first external gills; later internal gills; still later lungs
   (*b*) Through its skin; mucus membrane of the buccal cavity; and lungs

10. (*a*) Unisexual
    (*b*) Sexual reproduction
    (*c*) Mating occurs; the egg cells are released by the female in the water; and the male discharges sperms over it; i.e. fertilisation is external in the water

- (d) The eggs are protected by jelly-like albumen; against shock; and penetration by bacteria; embryo nourished by yolk in the egg
- (e) Larva or tadpole
- (f) Incomplete metamorphosis; it goes through various stages; to become an adult

11. (a) Reptilia
    (b) Dry soil; between or under rocks and leaves
    (c) 1. Nostril   2. Eye   3. Ear   4. Front leg   5. Hind leg   6. Cloaca
    (d) A. Head   B. Neck   C. Trunk   D. Tail
    (e) (i) Shields   (ii) Scales
    (f) S-shaped movements; of the trunk; limbs drag it across the ground; it operate diagonally and presses against the ground; in row-like movements

12. (a) Lungs
    (b) External intercostal muscles contract; the ribs are lifted; and the body cavity enlarges; resulting in the pressure on lungs decreasing; the lungs expand; and the air flows in

13. (a) Carnivorous; heterotrophic
    (b) The long tongue is shot out; tip of tongue is sticky; and the teeth which are outgrowths of the jaws; secure the prey; prey can be gripped few times to kill it; and it is swallowed whole

14. (a) Unisexual
    (b) Male and female mate; and sperm are transferred via cloaca to vagina of female; fertilisation is internal
    (c) Eggs are laid (oviparous, some ovoviviparous); amongst rotting leaves; the heat of the sun incubates eggs; there is no metamorphosis

15. (a) Aves
    (b) 1. Bill   2. Cere   3. Eye   4. Wing   5. Quill feathers   6. Hind leg
    (c) Scales   (d) Quill feathers   (e) 4. flying; 5. brakes when landing

16. The pigeon has large muscles; for depressing the wings; and a keel-like extension for attachment of muscles
    The skeleton is rigid but light; and air sacs reduce the specific gravity; some bones have large air cavities; the body is streamlined
    The wings with feathers form a large surface area for flight

17.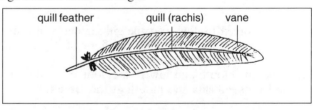

18. (a) Lungs and; air sacs
    (b) Air moves through lungs; into the air sacs; thus a constant stream of air flows into the lungs; and gaseous exchange occurs with inspiration and expiration

19. (a) Sexual   (b) Unisexual   (c) Internal (Fallopian tube)

20. (a) 1. Amniotic fluid   2. Embryo   3. Chorion
        4. Allantois   5. Air cavity   6. Calcareous shell
        7. Yolk   8. Yolk sac   9. Albumen
    (b) 1. Protects embryo against shock   4. Store wastes of the embryo
        7. Supplies food to the embryo

21. The body is divided into head; neck; trunk; and tail
    The body is covered with hair; with two pairs of legs; two ears with pinnae; two eyes with eyelids; and a nictitating membrane. There are also two nostrils; and mouth with fleshy lips
    The female has mammary glands. The anus is situated at the hind end. The sexual organs are the male testes; with penis; and the female vagina

22. (a) Internal; high up in Fallopian tubes
    (b) (i) Placenta
        (ii) They suckle their young; with milk; from mammary glands

HIGHER GRADE
QUESTIONS

Section A

A. Various possibilities are suggested as answers to the following questions. Indicate the correct answer.

1. Which of the following animals all have nictitating membranes?
   (i) Frog   (ii) Reptile   (iii) Fish   (iv) Insect   (v) Bird
   A (i), (ii) and (iv)   C (i), (ii) and (v)
   B (i), (iv) and (v)   D (ii), (iii) and (iv)

2. In the class Osteichthyes the following two organs are involved in depth control.
   A Eyes and swim bladder.   C Eyes and lateral line.
   B Swim bladder and ears.   D Lateral line and swim bladder.

3. Which one of the classes has the following characteristic: The eyes are on either side of the head and has no eyelids?
   A Osteichthyes   C Reptilia
   B Amphibia   D Aves

4. Which of the following represents the stroke of the caudal fin which propels the fish forward?
   A 1 and 2
   B 2 and 3
   C 1 and 3
   D 2 and 4

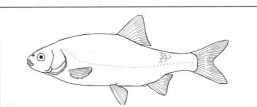

5. Water is forced to flow over the gill-filaments of a fish when the fish
   A closes its mouth and raises the floor of its mouth.
   B closes its gill-slits and opens its mouth.
   C open its mouth and lowers the floor of its throat.
   D open its gill-slits and mouth at the same time.

6. The characteristic which classifies this animal as Osteichthyes, is that it
   A has a naked skin.
   B has paired limbs.
   C has skin covered with scales.
   D is viviparous.

7. Three days after hatching, a tadpole obtains oxygen mainly through its
   A external gills.   C skin.
   B internal gills.   D lungs.

8. A newly hatched tadpole feeds on
   A the jelly from which it hatched.   C the leaves of water plants.
   B small animals in the water.   D the remains of egg yolk.

9. Which one of the following groups of characteristics are typical for the class Reptilia?
   A Hair, homeothermic, sweat glands, pinnae
   B Scales, ectothermic, shields, oviparous
   C Naked skin, ectothermic, cloaca, oviparous
   D Parental care, endothermic, cloaca, oviparous

10. Scales are found in
    A Osteichthyes and Reptilia only.
    B all Vertebrata.
    C Osteichthyes, Reptilia and Aves.
    D Osteichthyes only.

11. Internal fertilisation and the characteristic that they are oviparous are found in
    A Amphibia and Reptilia.
    B Mammalia and Aves.
    C Osteichthyes and Reptilia.
    D Aves and Reptilia.

12. The developing embryo of the pigeon excretes its wastes into the
    A yolk sac.
    B amnion.
    C chorion.
    D allantois.

13. Which of the following characteristics does **not** apply to birds?
    A Warm-blooded
    B Skin covering of feathers
    C Ovoviviparous
    D Legs covered with scales

14. The temperature of an endothermic animal is
    A high in warm weather.
    B low in cold weather.
    C high in warm weather and low in cold weather.
    D constant in any weather.

15. Select the phrase which is **not** related to all the others.
    A Homeothermic
    B Feathers
    C Scales
    D Diaphragm

16. The term "viviparous" means
    A feeding the young on milk.
    B producing offspring asexually.
    C producing offspring from unfertilised eggs.
    D producing young which have developed beyond the egg stage.

17. A characteristic only applicable to mammals is that they
    A are fertilised internally.
    B have external pinnae.
    C are homeothermic.
    D possess four limbs.

18. The young mammal developes in the
    A urethra.
    B ureter.
    C uterus.
    D vagina.

19. A mammal differs from all other Vertebrata in having
    A two pairs of pentadactyle limbs.
    B a constant body temperature.
    C a body covering of hair.
    D an air-filled middle ear.

20. Although fishes and birds are unisexual, birds differ from fishes in that birds
    A are oviparous.
    B are only externally fertilised.
    C are only internally fertilised.
    D ovoviviparous.

B. Supply the correct term for each of the following statements.

1. An embryonic membrane which acts as an organ of breathing in the embryo's of reptiles and birds
2. The subphylum of all animals with an endoskeleton
3. The type of reproduction where the eggs undergo little or no development before they are laid
4. A class of vertebrates, the members of which are endothermic have lungs but do not possesses ear pinnae
5. "White" of egg; main function is to supply embryo with water during its development
6. Protects the embryo's of birds, reptiles and mammals against shocks and injury
7. Wax-like naked skin that covers the base of a bill in birds
8. The chamber into which the rectum, genital ducts and the urinary ducts open in fishes, amphibians, reptiles and birds
9. Breathing organ of fishes
10. Transformation of a larva to an adult
11. The innermost, transparent eyelid found in birds and reptiles
12. A tissue complex, formed in part from the inner uterine lining and in part from the chorion of the embryo
13. The extra-embryonic membrane that encloses the yolk of the egg of a bird
14. The cell resulting from sexual fusion of two gametes

C. In which of the following animals

A. Fish   B. Frog   C. Lizard   D. Bird   E. Rat or Rabbit

1. is fertilisation external in the water?
2. is fertilisation internal?
3. is there no covering of the egg by jelly or a shell?
4. is there a protective albumen layer surrounding the egg?
5. is the egg surrounded by a shell?
6. is there oviparity?
7. is there viviparity?
8. and does the newly-hatched young not resemble the parent?

D. Give one function of each of the following:
1. Gill-rakers
2. Jelly-like albumen layer
3. Haemoglobin
4. Keel

E. Write down the letter of the description in column B which best suits the term in column A.

Column A		Column B
1. Tadpole	A	Characteristic of the class Aves
2. Ovoviviparous	B	Scales on head of lizard
3. Air sacs	C	Embryo develops in egg inside mother's body
4. Parental care	D	Increase specific gravity
5. Suckle	E	Keep skin moist for gaseous exchange
6. Osteichthyes	F	Breathe with external gills
7. Webs	G	On tail of lizard
8. Shields	H	Gaseous exchange in lungs
9. Mammalia	I	Skin covered with hair
10. Placenta	J	Help with gaseous exchange in the class Aves
	K	Scales on feet of pigeon
	L	Characteristics of the class Mammalia
	M	Possess no eyelids
	N	Characteristic of the class Amphibia

Section B

1. Answer the following questions on the pigeon's egg.

    (a) Name the class of vertebrates to which the pigeon belongs.
    (b) Identify the parts numbered 1 to 8.
    (c) State the functions of the parts numbered: (i) 6, and (ii) 8.
    (d) Mention the properties of the part numbered 4.
    (e) Write down the number of the part which provides the embryo during development with:
       (i) proteins and water; and
       (ii) carbohydrates and fats.
    (f) How long does it take for a pigeon's egg to hatch?

2. (a) Mention **four** characteristics which the classes Reptilia and Amphibia have in common.
   (b) Tabulate **six** important differences between the two groups mentioned in (a).

3. The bony fish obtains oxygen when water flows over the gills. One way of doing this is by moving the floor of the mouth up and down as shown below. Study the diagrams and answer the following questions.

   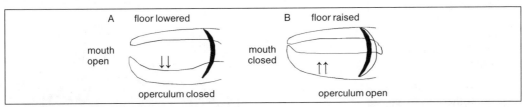

   (a) When the floor of the mouth is lowered, as in diagram A, what happens to the pressure inside the mouth?
   (b) When the mouth and gullet close and the floor of the mouth is raised, as in diagram B, what happens to the pressure inside the mouth?

141

(c) The gills of a living fish are deep red. In what way does this suggest that they are adapted to their function?
(d) The fish possesses 4 pairs of gills, each divided into many lamellae. In what way does this suggest that the gills are adapted to gaseous exchange?
(e) What is the function(s) of the operculum?
(f) Why is the gullet closed when the floor of the mouth is raised?

4. The table below shows the number of eggs produced by species of each of the five classes of vertebrates.

Class	Species	Average number of eggs produced
Mammalia	Baboon	1
Aves	Pigeon	2
Reptilia	Lizard	8
Amphibia	Frog	1400
Osteichthyes	Salmon	4 000 000

Answer the following questions:
(a) Which **two** examples mentioned above, produce eggs that are:
   (i) laid with shells and large quantities of yolk;
   (ii) fertilised internally, and
   (iii) fertilised externally?
(b) Relate the number of eggs laid with (ii) and (iii) above.
(c) The frog and salmon produce large numbers of eggs but the adult population remains fairly constant. Give an explanation for this phenomenon.
(d) Rabbits usually produce only a few eggs at a time. What features of the reproduction of rabbits improve the chances of:
   (i) the eggs being fertilised, and   (ii) the offspring being born?

5. A lizard were kept in an open enclosure in a laboratory. At regular intervals during a period of three days the rectal temperature of the lizard and the temperature of the enclosure were measured. The readings are shown in the graph. Study the graph and answer the questions which follow

(a) Is the lizard ectothermic or endothermic? Give a reason for your answer based on the graph.

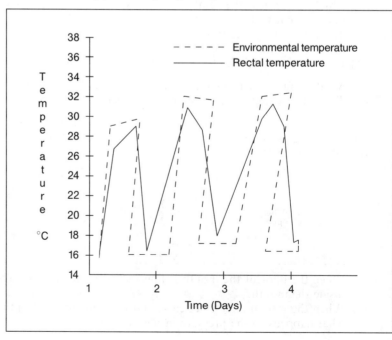

(b) Why did the lizard's body temperature fluctuate each day?
(c) Why did the lizard's body temperature never reached that of the environment?
(d) If a rat is kept under the same conditions, how would the graph of the rat's body temperature differ from that of the lizard's rectal temperature?

6. Study the following diagram of the reproductive cells in Amphibia.
   (a) Identify the parts numbered 1 to 6.
   (b) Identify the cells A and B.
   (c) Which process is taking place and where does this process usually take place?
   (d) Give the functions of numbers 3 and 4.
   (e) What organism develops from A within a few weeks?
   (f) Explain the gaseous exchange and nutrition of the organsim mentioned in (e).

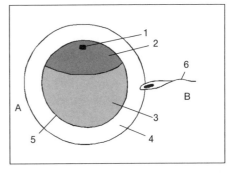

7. In tabular form, compare the classes Osteichthyes, Amphibia, Reptilia and Aves with regard to:
   (a) skin covering;
   (b) body division;
   (c) body appendages; and
   (d) locomotion.

8. The following diagram represents a section through the developing embryo of a bird.
   (a) Identify the membrane numbered 1. What fills the space between the embryo and this membrane?
   (b) Which number represents the yolk?
   (c) Which number represents the allantios. State **one** function of this structure.
   (d) What fills the space numbered 4?
   (e) Is the animal oviparous, ovoviviparous or viviparous?
   (f) Identify the part numbered 7. Mention two nutritive materials of this part.

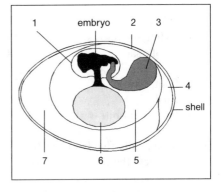

9. Explain why the mechanism of reproduction of the class Mammalia is considered to be more advanced than that of the class Osteichthyes.

10. What are the principal differences between the classes Amphibia and Reptilia which make the reptile better adapted for life on land?

11. Describe the main stages in the development of a frog from the fertilised zygote until it becomes an adult.

Section C

1. Discuss the reproduction of an earthworm and mammal referring to provision of
   (a) suitable sites for male gamete development;
   (b) adequate protection for the male gametes during their transfer to the female gametes; and
   (c) adequate nutrition for the developing embryo's.

2. Give a general account of the body covering of those classes of the Vertebrata you have studied and explain how the body covering of each is adapted to the specific mode of life of the animal.

3. Mention all the classes of the Vertebrata (except Mammalia) you have studied and discuss the adaptations of the adult individuals to:
   (a) locomotion,
   (b) gaseous exchange, and
   (c) nutrition.

HIGHER GRADE
ANSWERS

Section A

A. 1. C   2. D   3. A   4. D   5. A   6. C   7. A
      8. D   9. B   10. C 11. D 12. D 13. C 14. D
      15. D 16. D 17. B 18. C 19. C 20. C

B. 1. Allantois   2. Vertebrata   3. Oviparous
       4. Aves   5. Albumen   6. Amnion
       7. Cere   8. Cloaca   9. Gills
       10. Metamorphosis   11. Nictitating membrane 12. Placenta
       13. Yolk sac   14. Zygote

C. 1. Fish; frog   2. Lizard; bird; rat or rabbitt   3. Fish; rat or rabbitt
       4. Bird   5. Lizard; bird   6. Fish; frog; lizard; bird
       7. Rat or rabbitt   8. Frog

D. 1. Keeps gills free from particles which could block it
       2. Protects the egg of the bird
       3. Transports oxygen
       4. For the attachment of the pectoral muscles of the bird

E. 1. F   2. C   3. A   4. A/L   5. L
       6. M   7. N   8. B   9. I   10. L

Section B

1. (a) Aves
    (b) 1 – albumen   2 – yolk   3 – embryo
        4 – shell   5 – shell membrane   6 – air space
        7 – vitelline membrane   8 – chalaza
    (c) (i) Acts as shock absorber when egg is laid; allows expansion of egg content with increase in temperature during development
        (ii) Keeps yolk in position so that nucleus is directed uppermost
    (d) Calcareous and porous   (e) (i) 1 (ii) 2   (f) 20 days after start of hatching

2. (a) Four limbs; sexes separate; possess a cloaca; ectothermic; nictitating membrane
   (b) *Differences*

Reptilia	**Amphibia**
Scales and shields	Skin naked
Dry skin	Skin moist
Pentadactyl	Four fingers/five toes
Fingers and toes end in horny claws	No claws or nails
Breathing by lungs	Lungs, skin and mucus membrane of mouth
Internal fertilisation	External fertilisation
Head, trunk and tail	Head and trunk only
Movable eyelids	Eyelids immovable
No metamorphosis	Metamorphosis

3. (a) Pressure decreases; water flows into mouth
   (b) Pressure increases; water out through external gill-slits
   (c) Well supplied with blood capillaries; and blood; for absorption and transportation of respiratory gases
   (d) Large surface area for gaseous exchange
   (e) Protect gills; by moving in and out; in conjugation with the floor of the mouth
       It helps with ventilation of the gills
   (f) To prevent water from entering the alimentary canal

4. (a) (i) Aves – pigeon   (ii) Mammalia – baboon   (iii) Amphibia – frog
           Reptilia – lizard   Aves – pigeon   Osteichthyes – salmon
                                      Reptilia – lizard
   (b) Internal fertilisation; few eggs at a time; external fertilisation; large number of eggs laid
   (c) Large number of eggs laid; small eggs serve as food for predators; chances of survival thus small; the more eggs being laid; the better the chances of survival
   (d) (i) Eggs fertilised internally; sperm transported by penis of male; thus chances of survival excellent
       (ii) Embryo protected; inside uterus; of mother; and directly supplied with food; and respiratory gases; wastes removed by placenta; through umbilical cord

5. (a) Ectothermic; its temperature fluctuates; with that of the environment
   (b) Because the environmental temperature also fluctuates
   (c) Heat energy; produced by the lizard; is given off to the colder environment
   (d) Would be a straight line

6. (a) 1. Nucleus   2. Animal pole
       3. Vegetative pole with yolk   4. Gelatinous envelope of albumen
       5. Cell membrane   6. Flagellum
   (b) A – egg cell/ovum   B – sperm cell
   (c) Fertilisation; external in stagnant or slow-flowing water
   (d) 3 – provides embryo and young larva with food
       4 – protects egg when it is laid; acts as protective cushion; makes it difficult for birds to pick up; the eggs protects the embryo against harmful micro-organisms
   (e) Larva/tadpole
   (f) Exchange of gases initially by means of external gills; later internal gills
       It feeds initially on yolk of original egg cell; and afterwards on plant material

7.

	Osteichthyes	Amphibia	Reptilia	Aves
(a)	Slimy skin with scales	Moist skin	Scales and shields	Feathers and scales
(b)	Head, trunk and tail	Head and trunk	Head neck, trunk, and tail	Head, neck, trunk, and tail
(c)	Fins paired and unpaired	Two pairs of limbs	Two pairs of legs (except snakes)	Two wings and two legs
(d)	Swim	Jump and swim	Glide with sinuous movement	Walk and fly

8. (a) Amnion; watery fluid (b) 6
   (c) no. 3 – Serves as receptacle for nitrogenous wastes, e.g. uric acid
   (d) Air   (e) Oviparous   (f) Albumen; proteins and water

9. In mammals the genital ducts open directly to the exterior; while fishes have a cloaca
   Mammals have internal fertilisation; with less wastage of gametes
   The eggs do not have as much stored food; as yolk; because the embryo develops; protected by the mother in her uterus; and is fed by the placenta
   The small are more developed at birth than the fish; and less eggs are produced than the fish
   The maternal instinct is well developed and the young are suckled

10. *Reptiles*
    Have scales; and shields; and their skin is dry; for better protection on land than moist skin
    The limbs are modified for walking; more efficient than jumping
    They breathe with lungs only; therefore less chance of desiccation
    Fertilisation internal which is more effective
    The eggs are larger and laid on land with protective shell
    There is no water-living larval stage requiring gills for gaseous exchange

11. The zygote is formed after copulation; and the jelly surrounding the egg swells
    The zygote begins to divide; and hatches after ± 3 days
    It hangs on water weeds; and possesses no mouth; and feeds on yolk of egg
    The gills are external; for gaseous exchange; and the mouth formed after ± 5 days
    The external gills; are replaced by internal gills
    Lungs develop after ± 10 days; and tadpole comes to surface to breathe
    The hind limbs start to develop and forelimbs after ± 5 weeks
    The tadpole matures after ± 7 weeks;
    The tail is absorbed; and used as food; and it comes out of water as young adult

Section C

1. (a) **Suitable sites for formation of gametes**
   Earthworm:
   The earthworm has one pair testes; in segments ten and eleven; and is well protected inside of the body; the male ducts open on segment fifteen
   Mammal:
   The mammal has one pair of testes; outside the trunk in the scrotum; which has a lower temperature; than within body to produce sperm

   (b) **Protection of male gametes**
   Earthworm
   The gametes move in a seminal fluid; along seminal grooves; to spermathecae of the other worm; in the protective mucus layer; formed by special glands around both worms
   The sperms are stored in spermathecae; of the other worm; until the eggs are ripe

Mammal:
The penis transports sperms; into vagina
The seminal fluid allows sperms to swim; up in Fallopian tube

(c) **Nutrition of the embryo's**
Earthworm:
A small quantity nutritive liquid; and albumen; by special glands; supply nutrition
Mammal:
The chorion develops villi; into the blood-spaces; the placenta; with umbilical cord for nutrition of embryo
Small amount of yolk; and nutritive liquids of endometrium; serve as nutrition before implantation in uterus

2. The body covering of animals are adapted to their environment and the main function is that of protection
Osteichthyes
Horny scales; arranged like tiles on a roof; many mucus glands; streamline the fish for movement through water; and escaping predators
Amphibia
The skin is naked; thin; and moist; as well as glandular; with many blood vessels; for gaseous exchange; and pigment cells for camouflaging; against enemies
Some glands secrete toxic substances; for protection against enemies
Reptilia
The scales and shields are arranged like tiles of roof; with no glands
This protect animal against mechanical injury; and pigment cells for camouflaging
Moulting occurs
Aves
The feathers with their free ends backwards; and waterproofed by glands; keep the body temperature (endothermic).
Feathers make the bird streamlined; and enlarge surface of wings for flight
The scales on the feet are for protection
Mammalia
Covered by hair; with a layer of air trapped; for endothermic animal
Glands in skin; are for regulating body temperature; and excretion of metabolic wastes
Oil glands protect skin; against desiccation; and milk glands are present for nutrition of young

3. (a) **Adaptations with regard to locomotion**
Osteichthyes
The body is streamlined; and covered with overlapping scales; for less resistance to water
The unpaired fins; enlarge area from top to bottom; and the paired fins; serve as steer; and balance organs
The caudal fin; on the tail; and strong trunk; and tail muscles; propel the fish forward
Amphibia
The body is streamlined; with short front limbs; to absorb shock; when frog lands after a jump; and to steer when swimming
The hinds limbs are long; with strong muscles; for jumping; and swimming
Webbed toes; supply a large surface area; for resistance to water
Reptilia
Two pairs of short pentadactyle limbs; between which the trunk is suspended; are for locomotion
The neck; and tail; facilitate easy sinuous movement; and the five toes are for greater resistance to ground

Aves
The front limbs are modified into wings; for flight;
The body is covered with feathers; increasing the surface area of wings
The feet for walking; has claws for balancing
The palm and wrist bones of the wings are fused for rigidity
The skeleton is strong; and rigid; and light; so that the specific gravity of body is low
Some bones are hollow with air cavities
The body is streamlined; and feathers possess interlocked barbules; forming a firm surface

(b) **Adaptations with regard to gaseous exchange**
Osteichthyes
They breathe with special organs; namely gills in the water
The gill-rakers on the gill-arch; sieve out food from water; and prevent damage to the delicate gill-filaments
The gill-filaments are thin; and supplied with blood capillaries
The operculum and floor of the mouth; cause a pumping action; to force water over the gills
Amphibia
They breathe through skin; mucous membrane of mouth; and lungs
The skin is moist; as a result of glands in the skin; and supplied with blood capillaries; for diffusion of gases
The mucous membrane of mouth; also moist and supplied with capillaries
The gases are exchanged; by pumping action of the floor of the mouth
The two lungs with ridges; is to increase the gaseous exchange surface; it is also well supplied with capillaries
Their pumping action exchange air in the lungs
Reptilia
The respiratory organs are lungs; and they have nostrils to let air in
The inside of the lungs are supplied with capillaries; and has an inner area convoluted to increase the surface area
It is also moist; to facilitate gaseous exchange
The ribs with; intercostal muscles supply the respiratory movements
Aves
Breathing with lungs are aided by air sacs
The spongy lungs; with criss-cross bronchial tubes; lined with thin and moist membrane; supplies an efficient exchange surface
The abdominal and intercostal muscles; create the breathing movements; and air sacs create a constant stream of air during in- and exhalation

(c) **Adaptations regarding nutrition**
Osteichthyes
They are either herbivorous; or carnivorous
The carnivores have teeth; pointing backwards; preventing the prey from escaping
The prey is swallowed whole
Herbivores sieve algae; and other food with gill-rakers; from the water
Amphibia
The adult is carnivorous; and the tongue are attached to front; and flicked out to catch the prey; with its sticky tip which are then placed in mouth; to be swallowed
Reptilia
They are mainly carnivorous; and the prey is caught with a forked tongue
The prey is held by jaws; with teeth on both of them; and it is swallowed whole
Aves
They have no teeth; and the bill peck the food; which is swallowed whole
The shape of the bill is adapted for the type of food
Birds of prey have strong claws; to catch prey and for transport
Seed-eaters store food in a crop; where small stones grind it up

8 The role of the Nucleus and Cell division

It is recommended that students should first do this chapter on the role of the nucleus and cell division before studying the reproduction in bacteria and plants, and the section on Genetics.

A. The role of the nucleus
A structure characteristic of living cells is the **nucleus**.

1. Functions of the nucleus
(a) It controls the **structure** off a cell. It is involved in the production of ribosomes and RNA, substances needed for the synthesis of proteins.
(b) It controls the *activity* (metabolism) of the cell.
(c) It provides a mechanism, through cell division, for the transmission of **hereditary characteristics**.

2. Structure of the nucleus

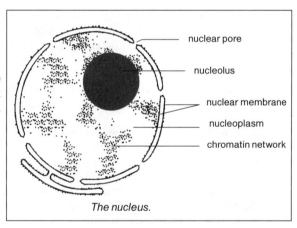

The nucleus.

Each nucleus is surrounded by a **double nuclear membrane** perforated by **pores**. It contains **nucleoplasm**, a **nucleolus** and chromatin network or **chromosomes**. Chromosomes contain the genetic material of a cell.

3. Nucleic acids
The two main kinds of nucleic acids are **DNA** and **RNA**. DNA is found mainly in the thread-like chromosomes in the nucleus. RNA is located mainly in the **nucleolus** and **cytoplasm**, particularly at the ribosomes.

(a) **The structure of nucleic acids** (HG only)
The nucleus contains **nucleic acids** formed from building blocks called **nucleotides**. The DNA molecule controls cell metabolism and holds hereditary information. The RNA molecule carries the hereditary information from DNA to the cytoplasm of the cell.
  A nucleotide consists of a **pentose sugar** (deoxyribose or ribose), a **phosphate group** and one of five possible **nitrogenous bases**, viz. adenine (A), guanine (G), cytosine (C), thymine (T) and uracil (U). DNA contains A, G, C and T; and RNA A, G, C and U.
  In a DNA molecule the nucleotides form two long chains which join across the nitrogenous bases by weak **hydrogen bonds**. DNA is coiled up in a **double helix** with the sugar groups

interconnected by pairs of bases linked in a specific way: A with T and C with G. A DNA molecule, which is enormously long, is a very important constituent of the **chromosomes** and consists of many shorter active units, called **genes**. A single gene codes for only one protein and is the unit of genetic information.

(b) **DNA replication** (HG only)
DNA has the ability to **replicate** itself accurately. In the cell, replication occurs during interphase, prior to cell division.

The two DNA strands uncoil gradually and separate as the hydrogen bonds break. At each nitrogenous base a **complementary nucleotide** is built on the orginal strand and two separate and identical double helices are formed.

The parent strand is **retained** or 'conserved' and each 'new' DNA molecule is really half-old, half-new.

This mode of DNA replication is called **semi-conservative** replication. After replication the DNA molecule assume a **circular form** to prevent unwanted replication. When upsets in the joining on of complementary bases occur, it results in **gene mutations**.

(c) **Functions of nucleic acids**
DNA • carries the hereditary characteristics
 • regulates the various processes of the cell
RNA • plays a role in protein synthesis

(d) **Role (functions) of nucleic acids in protein synthesis** (HG only)
Proteins are large complex molecules consisting of at least 20 different types of amino acids. The structure of a protein molecule is largely determined by the sequence in which the amino acids are joined to form the polymer chain.

DNA exerts its influence by controlling **protein synthesis**. It controls protein synthesis by determining the sequence in which amino acids are linked together on the ribosomes. Each amino acid is coded for by a triplet of nitrogenous bases in DNA (**codon**). The sequence of amino acids in the protein is determined by the sequence of base triplets in DNA.

In controlling protein synthesis, the DNA is first transcribed (transcription) into a **messenger RNA** (mRNA). When a mRNA is transcribed, it separates from its **DNA template** and is transferred through the intervention of **transfer RNA**, to the ribosomes in the cytoplasm. At the ribosomes the amino acids are fixed according to the code supplied by the triplets of nitrogenous bases.

The four main stages in protein synthesis are:
(a) DNA provides the **master plan** in code form.
(b) Transcription, that is the synthesis of mRNA.
(c) Activation of amino acids to combine it with tRNA.
(d) Translation, that is the means by which a specific sequence of amino acids is formed along the mRNA molecule and includes **initiation** of anticodon of tRNA by a polysome, **chain elongation** and **chain termination**.

B. Cell division

1. **Cell division** is responsible for **growth** and **reproduction**. In both cases the chromosomes must be correctly distributed between the daughter cells. There are two types of cell division, namely **mitosis** and **meiosis**. **Mitosis** is involved in the division of somatic cells to make **exact copies** of the nucleus of the original cell for growth and repair of tissues.

   Meiosis is reduction division and is involved principally in the formation of **gametes** for sexual reproduction and in certain plants **spores** for asexual reproduction. Normally a cell contains two of each type of chromosomes – the **diploid stage**. Mitosis preserves this condition but meiosis results in four daughter cells containing only one of each type of chromosome – the **haploid stage**.

2. Mitosis and meiosis have two steps, namely **karyokinesis** (nuclear division) and **cytokinesis** (cytoplasmic cleavage). Both types of cell division involve the orderly movement of chromosomes on a **spindle apparatus**.

3. Cell division can be conveniently divided into four phases which merge into one another with interphase as an intermediate phase.

   In **interphase** the cell prepares for division and the genetic material replicates so that there are two copies of each DNA molecule, one for each of the two daughter cells, to be produced.

   In **prophase** the chromosomes make their appearance as distinct bodies, the nucleolus fades and the nuclear membrane disintegrates.

   In **metaphase** the chromosomes arrange themselves at the centre of the spindle equator. They become attached to the spindle fibres at the centromeres.

   In **anaphase** the chromatids migrate to opposite poles of the spindle. The energy for this contraction is provided by ATP.

   In **telophase** the chromatids arrived at opposite spindle poles and are now called daughter chromosomes. The chromosomes uncoil and decondense in thread-like form. The spindle fibres disintegrate and a nuclear membrane is formed.

4. In **mitosis** the **homologous chromosomes** do not associate with one another. During metaphase they arrange themselves independently on the spindle equator and during anaphase each **centromere** divides into two equal parts, the two chromatids separate and migrate to opposite poles independently from one another. In this way the diploid state is preserved.

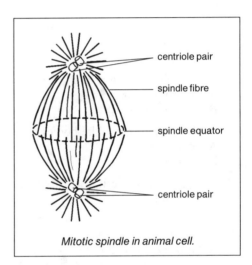

Mitotic spindle in animal cell.

5. In **meiosis** the nuclear division involves two successive divisions. In the first meiotic division the number of chromosomes is reduced (halved) and genetic material is exchange at certain points called **chiasmata**. The exchange of portions of chromatids is called **crossing over**. The second meiotic division proceeds like a normal mitotic division and each of the four daughter cells formed, has half the number of chromosomes and is different in genetic make up.

6. The **significance of mitosis** can be summarised as follows:
   (i) It is the process which allows multicellular organisms to grow in size.
   (ii) The daugther cells are both structurally and functionally like the parent cell. They contain exactly the same number of chromosomes with exactly the same genetic material (DNA) as the parent cell.
   (iii) It allows for the replacement of worn out and repair of damaged cells and tissues.
   (iv) Most unicellular organisms reproduce asexually by mitosis.
   (v) It is responsible for the growth and development of structures, e.g. cuttings, which reproduce vegetatively.

7. The **significance of meiosis** can be summarised as follows:
   (i) Since **reduction in the number of chromosomes** occurs, meiosis ensures that the number of chromosomes in the cell of successive generations will remain constant.
   (ii) Each cell has a unique combination of genes which ensures variability in characteristics in the offspring. The possibility of **genetic variation** is increased by crossing over at chiasmata and **random assortment** which occurs during metaphase I.
   (iii) The daughter cells formed during meiosis differentiate into gametes. In plants the sporophyte produces haploid spores by meiosis which give rise to the gametophyte. Meiosis, therefore, makes **alternation of generations** possible.

STANDARD GRADE
QUESTIONS

Section A

A. Various possibilities are suggested as answers to the following questions. Indicate the correct answer.

1. The control centre of a living cell is the
   A   nucleolus.
   B   chromosomes.
   C   nucleus.
   D   ribosomes.

2. The main component of the nucleolus is
   A   RNA.
   B   DNA.
   C   protein.
   D   the chromatin network.

3. The small spherical bodies within the cell nucleus not bounded by a membrane, are
   A   chromosomes.
   B   nucleoli.
   C   nuclear pores.
   D   ribosomes.

4. Which of the following is found between the double layers of the nuclear membrane?
   A   Ribosomes
   B   Vacuoles
   C   DNA and RNA
   D   Perinuclear space

5. The primary function of the cell nucleus is
   A   reproduction.
   B   cell functioning.
   C   respiration.
   D   storage.

6. Dark-staining threadlike bodies within a cell nucleus are known as
   A   chromosomes.
   B   ribosomes.
   C   nucleoli.
   D   centromeres.

7. The proteins of a cell are synthesised on organelles called
   A   plastids.
   B   ribosomes.
   C   chromatids.
   D   chromosomes.

8. The centrosome
   A   attaches two chromatids together.
   B   plays a role in cell division in plant cells.
   C   contains pigments.
   D   plays a role in cell division in animal cells only.

9. This cell undergoes mitosis because
   A   chromosomes are arranged in a single line on the spindle equator.
   B   chromosomes are arranged in homologous pairs.
   C   crossing-over is taking place.
   D   chromosomes are single-threaded.

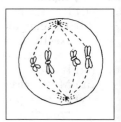

10. Meiosis is the process during which
    A   2 daughter cells identical to the parent cell are formed.
    B   4 daughter cells identical to the parent cell are formed.
    C   the chromosome number remains unchanged.
    D   the diploid chromosome number is reduced to the haploid number.

11. Anaphase is a stage in mitosis which is characterised by the
    A   formation of two daughter cells.
    B   movement of daughter chromosomes to opposite poles.
    C   formation of a spindle.
    D   arrangement of chromosomes on the spindle equator.

12. In a zygote the chromosome number is
    A diploid.
    B 15.
    C half the number present in a gamete.
    D haploid.

13. Which phase of mitosis is illustrated by this diagram?
    A Telophase
    B Prophase
    C Metaphase
    D Anaphase

14. A certain cell's nucleus has 12 chromosomes. How many chromosomes will each new cell nucleus have after mitosis?
    A 3
    B 6
    C 12
    D 24

15. Daughter cells formed by a cell with 20 chromosomes will each have
    A 10 chromosomes after mitosis.
    B 20 chromosomes after meiosis.
    C 40 chromosomes after mitosis.
    D 10 chromosomes after meiosis.

16. All of the following are phases of mitosis **except**
    A prophase.
    B interphase.
    C anaphase.
    D metaphase.

17. Mitosis ensures that each new cell produced by cell division will have
    A a full set of chromosomes.
    B an equal share of cytoplasm.
    C twice the number of chromosomes.
    D half the number of chromosomes.

18. Which of the following pairings is **not** true for mitosis?
    A Metaphase – chromosomes assemble at spindle equator
    B Anaphase – chromatids separate
    C Prophase – spindle forms
    D Telophase – chromosomes coil up

19. Spindle fibres in cell division do **not** connect
    A daughter chromatids to each other.
    B chromosomes to centrioles.
    C centromeres to centrioles.
    D centrioles to centrioles.

20. The duplication of chromosomes and division of a cell nucleus to form two nuclei, each with a complete set of chromosomes, is called
    A DNA replication.
    B meiosis I.
    C mitosis.
    D cell division.

Questions 21 to 25 refer to the following diagrams of two phases in mitosis.

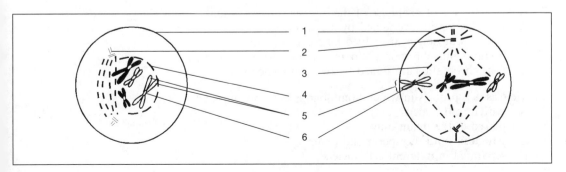

21. Which divided structures first move apart during mitosis?
    A 2   C 4
    B 3   D 5

22. How many chromatids are shown in the cell?
    A 2   C 6
    B 4   D 8

23. Which structure disappears when structure 3 is forming?
    A 1   C 3
    B 2   D 4

24. After completion of this division, each new cell's nucleus will have
    A 2 chromosomes.   C 4 chromosomes.
    B 2 chromatids.   D 8 chromosomes.

25. Which structure divides last as the chromosomes are about to migrate to opposite poles of the cell?
    A 1   C 5
    B 4   D 6

26. Which of the following cells in angiosperms have homologous chromosomes?
    A Sperms   C Endosperm mother cells
    B Microspores   D Gametes

27. The term that describes nuclear division is known as
    A gametogenesis.   C cyclosis.
    B karyokinesis.   D cytokinesis.

28. A cell that contains 15 chromosomes in the nucleus is
    A a zygote.   C a gamete.
    B diploid.   D impossible.

29. The correct sequence of the phases of mitosis is
    A metaphase → prophase → anaphase → telophase.
    B prophase → anaphase → metaphase → telophase.
    C telophase → prophase → anaphase → metaphase.
    D prophase → metaphase → anaphase → telophase.

30. Replication of the DNA molecule takes place during
    A interphase of mitosis.   C telophase of meiosis.
    B telophase of mitosis.   D metaphase.

31. When a diploid cell divides by meiosis the result will be
    A 2 diploid cells.   C 2 haploid and 2 diploid cells.
    B 4 haploid cells.   D 4 diploid cells.

32. A characteristic of prophase of the mitosis process is that
    A two chromatids are joined by a centromere.
    B two chromosomes are joined by a centromere.
    C homologous chromosome pairs separate from each other.
    D four chromatids are joined by two centromeres.

33. Meiosis is a process in which homologous chromosomes are
    A paired in the sperm only.
    B paired in the ovum only.
    C paired in both the sperm and ovum.
    D separated in different cell nuclei.

34. The product of meiosis in a diploid cell is
   A four identical cells.
   B four haploid cells.
   C two different diploid cells.
   D four cells having the same chromosome number as parent cell.

35. Replication of DNA takes place
   A during the metaphase of meiosis.
   B during protein synthesis.
   C during the interphase of mitosis.
   D between meiosis I and meiosis II.

36. Each somatic cell in the human possesses
   A 46 similar chromosomes.
   B 23 different chromosomes.
   C 46 pairs of different chromosomes.
   D 23 pairs of chromosomes.

B. Write down the correct term for each of the following statements.

1. The group of compounds to which DNA and RNA belong
2. A small RNA-containing body within the cell nucleus
3. A linear arrangement of genes
4. Carriers of genetic characteristics in nuclei
5. The nucleic acid which mainly occurs in the cytoplasm
6. The stage in the cycle of a cell when neither the nucleus nor the cytoplasm is actively dividing
7. Gene-containing thread-like body which appears conspicuously during cell division in the nucleus
8. One half of a chromosome as seen during cell division
9. The phase in cell division during which the chromosomes migrate to the spindle poles
10. The point at which homologous chromosomes join and crossing-over occurs in meiosis
11. A chromosome number twice that of a gamete of a given species
12. The individual chromosome threads
13. The process by means of which gametes are formed during meiosis
14. The change in chromosome number form 2n to n
15. The process during which reduction of chromosomes takes place
16. The phase in mitosis when the chromosomes arrange themselves on the spindle equator in such a way that the centromeres lie on the equatorial plane
17. The nuclear division when the chromosomes during prophase arrange themselves on the spindle equator in such a way that the centromeres lie on either side of the equatorial plane
18. The organelle which gives rise to the spindle in a dividing cell
19. The phase during mitosis in which the chromosomes arrange themselves in a plane centrally placed in the cell, at right angles to the spindle axis
20. The mutual exchange of genetic information between homologous chromosomes at the chiasmata
21. The region on a chromosome at which a spindle fibre is attached during cell division
22. The cleavage of the cell cytoplasm which follows nuclear division
23. The process during which the cytoplasm divides in two after nuclear division

C. Write down the letter of the description in column B which best suits the term in column A

Column A		Column B
(a) 1. Nucleolus	A	Forms spindle during cell division
2. Centriole	B	Controls functioning and structure of cell
3. Ribosome	C	Joins chromatias together
4. Centromere	D	Controls production of ribosomal RNA
5. Nucleus	E	Plays role in protein synthesis

Column A		Column B
(b) 1. Anaphase	A	Nuclear membrane disintegrates and spindle is fully formed
2. Prophase	B	Daughter chromosomes separate and move to opposite poles
3. Interphase	C	Chromosomes arrange themselves on spindle equator
4. Telophase	D	Neither the nucleus nor the cytoplasm is actively dividing
5. Metaphase	E	Chromosomes uncoil, lengthen and become invisible

Section B

1. Study the diagram of a nucleus and answer the questions.

    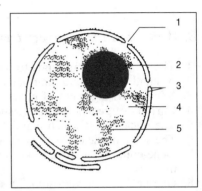

    (a) Write down the **number** and **name** of the part where each of the following takes place:
      (i) storage of genetic information;
      (ii) control production of ribosomal RNA.
    (b) Identify the structure numbered 3. Of which organic substances is this structure composed?
    (c) Identify the opening numbered 1.
    (d) State **three** functions of the nucleus.

2. (a) Name two important nucleic acids found in a living cell.
   (b) State the functions of one of the nucleic acids mentioned in (a).

3. The simplified diagram shows the nucleus of a somatic cell of a certain organism.

    (a) (i) Draw diagrams to show the **two** nuclei been formed if the cell divides mitotically.
        (ii) What is the biological significance of this type of nuclear division?

    (b) (i) Draw diagrams to show **four** different nuclei which could be formed if the cell divided by meiosis assuming that no crossing-over has taken place.
        (ii) What is the biological significance of this type of nuclear division?

4. The following diagrams represent phases in a cell division process.

    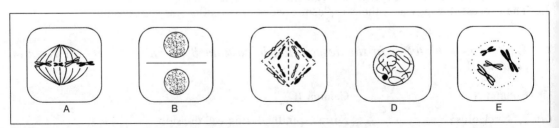

    (a) What type of cell division is illustrated here? Give **three** reasons for your answer.
    (b) Do these diagrams represent plant cells or animal cells? Give **two** reasons for your answer.
    (c) Identify the phases numbered A to E.
    (d) Arrange the phases mentioned in (c) in the **correct sequence**, starting with the interphase.

5. The following diagram represents a phase in a nuclear division process.

   (a) What type of nuclear division is illustrated here? Give **three** reasons for your answer.
   (b) What phase is represented here?
   (c) What will be the chromosome number in each cell as a result of this nuclear division?
   (d) Name **one** place where you would expect to find such division taking place in
       (i) an angiosperm, and
       (ii) the human.
   (e) Give one word to describe the paired chromosomes numbered A.
   (f) What process is taking place at B?

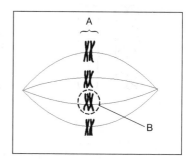

6. The following diagram represents a phase in nuclear division taking place in a somatic cell of a certain animal.

   (a) What type of nuclear division is illustrated here? Give **two** reasons for your answer.
   (b) What phase is represented here?
   (c) Identify the parts numbered 1 to 6.
   (d) How many chromosomes
       (i) are present in this cell,
       (ii) would be present in each cell after this division is completed?

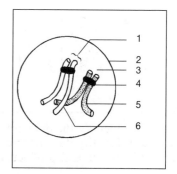

7. Describe the prophase stage of mitotic division in an animal cell.

8. Study the diagram on cells and answer the questions.

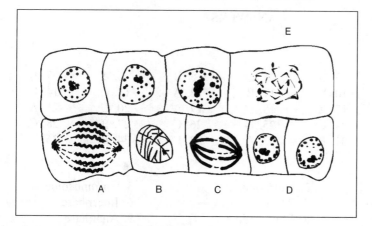

   (a) What process is depicted here?
   (b) Does the diagram represents plant cells or animal cells?
   (c) Identify the phases numbered A to E.
   (d) Briefly describe what is happening in each of the cells A and C.

9. Tabulate **eight** differences between mitosis and meiosis I.

10. The diagrams below represent two chromosomes during a stage in the division of a cell.

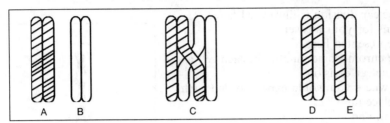

(a) During which type of cell division does this process, shown here, occur?
(b) What is the pair of homologous chromosomes, indicated by A and B, called?
(c) What process is taking place in C?
(d) Of what significance is the process, named in C, to the organism concerned?

11. Study the following diagrams and answer the questions.

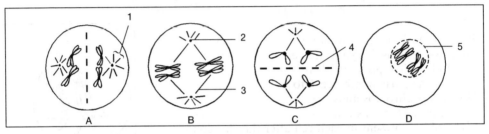

(a) Identify the different meiotic phases A to D.
(b) Identify the parts numbered 1 to 5.
(c) Which end products are formed as a result of this division?

STANDARD GRADE
ANSWERS

Section A
A. 1. C   2. A   3. B   4. D   5. B   6. A   7. B
    8. D   9. A   10. D   11. B   12. A   13. C   14. C
    15. D   16. B   17. A   18. D   19. A   20. C   21. A
    22. D   23. D   24. C   25. D   26. C   27. B   28. C
    29. D   30. A   31. B   32. A   33. D   34. B   35. C
    36. D

B. 1. Nucleic acids   2. Nucleolus   3. Chromosome
    4. Chromsomes   5. RNA   6. Interphase
    7. Chromosome   8. Chromatid   9. Anaphase
    10. Chiasma   11. Diploid   12. Chromatids
    13. Gametogenesis   14. Meiosis   15. Meiosis
    16. Metaphase   17. Meiosis   18. Centriole
    19. Metaphase   20. Crossing-over   21. Centromere
    22. Cytokinesis   23. Cytokinesis

C. (a) 1. D   2. A   3. E   4. C   5. B
   (b) 1. B   2. A   3. D   4. E   5. C

Section B

1. (a) (i) 5 – chromatin network   (ii) 2 – nucleolus
   (b) Nuclear nembrane • proteins and phospholipids
   (c) Nuclear pore
   (d) Regulates the metabolism of the cell
       Regulates the structure of the cell
       Carries the hereditary characteristics (genetic information)

2. (a) DNA and RNA
   (b) DNA • controls the processes of the cell; controls the synthesis of proteins

3. (a) (i)

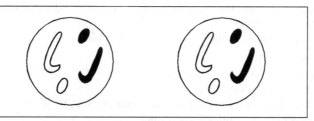

   (ii) Allows for multicellular organisms to grow in size
        Daughter cells are structurally and functionally like parent cell
        Daughter cells contain the same chromosome number and genetic material (DNA) as parent cell
        Allows for replacement of worn-out cells and tissues
        Allows for repair of damaged cells and tissues
        Unicellular organisms reproduce asexually by mitosis
        Allows for growth and development of structures which reproduce vegetatively

   (b) (i)

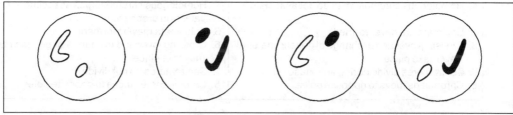

   (ii) Ensures that chromosome number in successive generations will remain constant
        Possibility of genetic variation increases by crossing-over at chiasmata and random assortment during metaphase I
        Makes alternation of generations in plants possible

4. (a) Mitosis • two daughter cells formed and not four
               • no crossing-over occurs or chiasmata formed
               • double stranded chromosomes at spindle equator and no bivalents
   (b) Plant cells • no centrosome with centrioles present
                   • cell plate is formed and no constriction of cell membrane
   (c) A – metaphase B – telophase C – anaphase D – interphase E – prophase
   (d) Interphase   Prophase   Metaphase   Anaphase   Telophase

5. (a) Meiosis • homologous chromosomes (bivalents) are present
   • chromosomes arranged on either side of equatorial plane
   • chiasma is present
   (b) Metaphase I   (c) Four (haploid)
   (d) (i) anther/ovule (ii) Ovary/testis
   (e) Bivalent   (f) Crossing-over

6. (a) Meiosis • homologous chromosomes (bivalents) are present
   • chiasma is found for crossing-over to take place
   (b) Prophase I
   (c) 1 – chromosome 2 – nuclear membrane 3 – nucleoplasm
   4 – centromere 5 – chromatid 6 – chiasma
   (d) (i) Two   (ii) One

7. DNA threads coil up and becomes shorter and thicker
   Visible as chromosomes and chromatids join at centromeres
   Nucleoplasm changes into gel state
   Nucleolus and mitochondria disappear and nuclear membrane disintegrates
   Centrioles divide and form poles
   Microtubules of spindle extend conically from pole to pole

8. (a) Mitosis   (b) Plant cells
   (c) A – metaphase B – prophase C – anaphase
       D – telophase E – interphase
   (d) A • In **metaphase** the spindle is fully formed and extends to poles
       • chromosomes arranged on spindle equator
       • chromatids become connected to spindle fibres at centromeres.
       C • In **anaphase** the centromere of each chromosome divides in two
       • chromatids are separated and spindle fibres contract
       • daughter chromsomes migrate to opposite poles.

9.

	Mitosis	Meiosis I
	1. The chromsome number remains the same	1. The chromosome number is halved
	2. Takes place in somatic cells	2. Takes place in gonads/sex organs
	3. During metaphase chromosomes arrange on spindle equator as separate units	3. During metaphase 1 chromosomes arrange in pairs on either side of spindle equator
	4. Homologous chromosomes do not associate	4. Homologous chromosomes associate to form bi-valents in prophase I
	5. Chiamata are never formed	5. Chiasmata may be formed
	6. Crossing-over and exchange of genetic material never take place	6. Crossing-over and exchange of genetic material may take place
	7. Centromeres divide during anaphase	7. Centromeres do not divide
	8. Chromatids move to opposite poles	8. Chromosomes move to opposite poles

10. (a) Meiosis (b) Bivalent (c) Crossing-over
    (d) Cross-links or chiasmata develop between homologous/chromosomes
        Homologous chromosomes exchange equivalent portions of genes
        New genetic combinations are produced and linked genes separated
        Homologous chromosomes end up in different gametes
        Genetic variation is increased
        Ensures that species constantly change and adapt when existing conditions alter.

11. (a) A – telophase   B – metaphase   C – anaphase   D – prophase
    (b) 1 – aster 2 – centriole 3 – spindle fibre 4 – spindle equator 5 – nuclear membrane
    (c) Four gametes; with two (n or haploid number) chromosomes in each

HIGHER GRADE
QUESTIONS

Section A

A. *Various possibilities are suggested as answer to the following questions. Indicate the correct answer.*

1. Which of the following is **not** a nitrogenous base of DNA?
   A Thymine
   B Cytosine
   C Guanine
   D Uracil

2. Which of the following is **not** found in DNA?
   A Ribose
   B Thymine
   C Deoxyribose
   D Cytosine

3. Which of the following carbohydrates forms a monomer of DNA?
   A Glucose
   B Maltose
   C Deoxyribose
   D Fructose

4. Which of the following describe nucleic acids and nucleotides the best?
   A Nucleic acids are monomers of nucleotides
   B Nucleic acids are acids and nucleotides are bases
   C Nucleotides are monomers of nucleic acids
   D Nucleotides are large molecules and nucleic acids small molecules

5. The configuration of base pairing in the DNA molecule is
   A C = T, A = G.
   B T = G, C = A.
   C A = G, T = U.
   D C = G, A = T.

6. The DNA of one species differs from others in its
   A sugars.
   B base-pair sequence.
   C site of production
   D phosphate groups.

7. When DNA replication commences
   A old strands move to find new strands before bonding.
   B the two strands of the double helix unwind.
   C the two strands condense tightly for transfer of nitrogenous bases.
   D two DNA molecules combine.

8. DNA replication produces
   A two double-stranded molecules totally different from parent molecule.
   B two double-stranded molecules, one with the old strands and one with newly assembled strands.
   C one double-stranded molecule genetically the same as parent molecule.
   D two half-old, half-new double-stranded molecules.

9. Where are the molecules of DNA most likely to separate during DNA replication?
   A Cytosine and guanine
   B Ribose and adenine
   C Phosphate ion and deoxyribose
   D Thymine and guanine

10. Ribosomes consist of
    A RNA and proteins.
    B RNA and deoxyribose.
    C DNA and proteins.
    D DNA an RNA.

11. Which of the following statements is true of RNA?
    A It is always a double helix of nucleotides
    B It is wound up as a helix
    C It contains uracil
    D It has a longer chain than DNA

12. During the replication of DNA, thymine always joins up with
    A   cytosine.
    B   a nucleotide.
    C   guanine.
    D   adenine.

13. If the sequence of bases in a region of a single DNA strand is:
    thymine – cytosine – thymine – guanine, the sequence of bases in the complementary strand will be
    A   thymine – cytosine – thymine – guanine.
    B   adenine – guanine – adenine – cytosine.
    C   cytosine – thymine – guanine – thymine.
    D   adenine – thymine – uracil – guanine.

14. A DNA nucleotide consists of.
    A   a phosphate ion plus ribose plus a base.
    B   deoxyribose plus a base.
    C   a phosphate ion plus a base.
    D   a sugar molecule plus a base plus a phosphate ion.

15. The purine bases of the DNA molecule are
    A   adenine and thymine
    B   cytosine and uracil.
    C   guanine and adenine
    D   guanine and thymine.

16. What is the complementary DNA sequence for the sequence AGT?
    A   AGT
    B   TCA
    C   UCA
    D   CTG

17. DNA replicates by breaking the bonds between its two strands, after which each strand
    A   synthesises a new strand.
    B   coils back upon itself.
    C   grows to double length.
    D   takes a spiral appearance.

18. The coded material that migrates from the cell nucleus to a ribosome where a particular polypeptide will be synthesised, is called
    A   ribosomal RNA.
    B   codon DNA.
    C   messenger RNA.
    D   transfer RNA.

19. Which DNA strand acts as a template during DNA replication?
    A   Both strands of DNA
    B   The one with all four nitrogenous bases
    C   The first one to receive a nucleotide
    D   The eldest of the two strands

20. Which is the most unstable RNA?
    A   All RNA is of equal stability
    B   mRNA
    C   tRNA
    D   rRNA

21. The transcription of TGA, codes for
    A   ACA in DNA.
    B   TGA in RNA.
    C   ACU in DNA.
    D   ACU in RNA.

22. Transfer RNA transfers
    A   polypeptides.
    B   amino acids.
    C   messenger RNA.
    D   anticodons.

23. All types of RNA originate from
    A   DNA.
    B   amino acids.
    C   nucleotides.
    D   uracil.

24. The first step in protein synthesis, where genetic information in DNA is transferred to RNA, is
    A   translation.
    B   amino acid activation.
    C   transcription.
    D   replication.

25. The RNA molecule is
    A a double helix.
    B usually single-stranded.
    C always double-stranded.
    D usually double-stranded.

26. The exact copying of the information stored as codons in DNA, onto a messenger RNA molecule take place
    A in the cytoplasm.
    B after translation.
    C in the nucleus.
    D on ribosomes.

27. The anticodon in tRNA for AAA in DNA molecule is
    A UUU.
    B TTT.
    C UAT.
    D AAA.

28. The coded genetic instructions for forming polypeptide chains are carried to the ribosomes by
    A rRNA.
    B DNA.
    C mRNA.
    D tRNA.

29. An anticodon pairs with the nitrogenous bases of
    A mRNA codon.
    B tRNA anticodon.
    C DNA codon.
    D amino acids.

30. The function of tRNA is to
    A pick up genetic messages from rRNA.
    B deliver amino acids to the ribosome.
    C construct mRNA on a DNA template.
    D synthesise mRNA.

31. Translation of a mRNA molcule with 48 nucleotides produces a polypeptide of
    A 3 amino acids.
    B 12 amino acids.
    C 16 amino acids.
    D 48 amino acids.

32. A gene mutation in codon 15 in a series of 20 codons would result in a polypeptide
    A of 14 amino acids only.
    B of 15 amino acids only.
    C with a different amino acid in position 15.
    D with the same structure as the original.

33. The number of different tRNA molecules a living cell at least must contain, is
    A 3.
    B 9.
    C 16.
    D 20.

34. A gene is that part of the genome that produces
    A various nucleotides.
    B one specific polypeptides.
    C one specific codon.
    D various polypeptides.

35. Nucleic acids consist of hydrogen bonded
    A deoxyribose
    B nucleotides.
    C amino acids.
    D phosphate ions and sugars.

36. ... starts when two ribosomal subunits, an initiator tRNA and a mRNA transcript, come together.
    A Translation
    B Replication
    C Amino acid activation
    D Transcription

37. Why must a codon consists of three nucleotides rather than two nucleotides?
    A Two nucleotides can not code for 20 different amino acids
    B There are three types of RNA molecules
    C There are three steps in protein synthesis
    D Protein synthesis can now proceed at a faster rate

38. X represents part of a strand of a DNA molecule

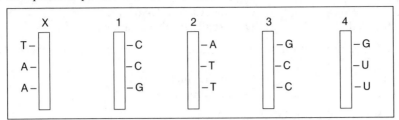

Which of the parts indicated as 1, 2, 3 or 4 is formed as a complimentary part to X during the replication of the DNA molecule?
A   1
B   2.
C   3.
D   4.

39. DNA and RNA are similar in that both molecules contain
A   ribose.
B   deoxyribose.
C   adenine.
D   uracil.

40. Which of the following combinations of bases is represented by X in the following DNA molecule?

A   GAC
B   GTC
C   AGC
D   CAG

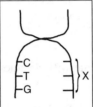

41. If the amount of DNA present in a cell at metaphase is 20 units, how much will be present in each nucleus immediately following telophase?
A   Zero
B   10 units
C   20 units
D   40 units

42. Interphase is the stage when
A   nothing occurs.
B   a dividing cell forms its spindle apparatus.
C   cytokinesis occurs.
D   a cell grows and duplicates its DNA.

43. Cytokinesis is a term that describes
A   nuclear division.
B   reducing the chromosome number.
C   cytoplasmic division
D   doubling the chromosome number.

44. The phase in the cell cycle in which the nucleus is mechanically inactive but chemically very active is
A   telophase.
B   interphase.
C   metaphase.
D   anophase.

45. Mitosis produces a daughter cell with genetic information. That is ... and with a chromosome number that is ... the parent cell.
A   rearranged; the same as
B   identical to parent's cells; one-half
C   rearranged; one-half
D   identical to parent's cells; the same as

46. A duplicated chromosome has
A   no chromatids.
B   one chromatid.
C   two chromatids.
D   four chromatids.

47. Mitosis takes place in
    A xylem elements.
    B the vegetative cells of algae during favourable growth conditions.
    C all plant and animal cells.
    D the cells producing gametes.

48. After completion of this division each daughter cell will possess the following number of chromosomes:
    A 8 as a result of mitosis.
    B 4 as a result of meiosis.
    C 2 as a result of mitosis.
    D 4 as a result of mitosis.

49. The process of mitosis occurs
    A during gametogenesis.
    B when a somatic cell divides.
    C in the gonads only.
    D when reduction division takes place.

50. Replication of chromosomes occurs during
    A crossing-over.
    B anaphase of mitosis.
    C metaphase II.
    D interphase.

51. Sexual reproduction without meiosis
    A is the general method used by organisms to survive.
    B occurs in lower plant types only.
    C would double the number of chromosomes each generation.
    D characterises the diploid life cycle.

52. A difference between the mitotic cell division in plant cells and that in animal cells, is the way in which the
    A cytoplasm divides.
    B chromosomes arrange on spindle equator.
    C genetic material replicates.
    D centromere divides.

53. Which phase of meiosis would show a chromosomal arrangement identical to a phase of mitosis for a cell with half as many chromosomes?
    A Metaphase I
    B Anaphase I
    C Metaphase II
    D Anaphase II

54. Nucleoli are not visible during mitosis because
    A ribosome transcription is taking place.
    B they are coiled up around histones.
    C they migrate to opposite poles.
    D they disintegrate like the nuclear membrane

55. The microtubules in a mitotic spindle stretch from
    A pole to pole.
    B pole to equator.
    C pole to chromosome.
    D centromere to centromere.

56. The stage when the nucleus is not in the process of dividing, is
    A telophase.
    B prophase I.
    C interphase.
    D metaphase I.

57. Mitosis is always responsible for the formation of
    A daughter cells with the haploid chromosome number.
    B daughter cells with the diploid chromosome number.
    C cells in which the number of chromosomes remain constant.
    D cells with new hereditary characteristics.

58. A characteristic of prophase of mitosis is that
   A two chromosomes are joined by a centromere.
   B two chromatids are joined by a centromere.
   C the chromosomes form homologous pairs.
   D four chromatids are joined by two centromeres.

59. Cells formed during the telephase of mitosis, differ from that of the prophase in that the
   A chromosome number in telophase is always haploid and in prophase diploid.
   B prophase cell contains double the amount of genetic material.
   C number of centromeres in the two phases differ.
   D genetic information of the two cells differs.

60. DNA and RNA are similar in that both nucleic acids
   A occur only in the nuclei of cells.
   B contain the nitrogenous base thymine.
   C contain the nitrogenous base guanine.
   D contain the sugar ribose.

B. Write down the correct term for each of the following statements.

1. The monomer of nucleic acids
2. The base of the DNA molecule which pairs with thymine
3. The process which gives rise to a gamete containing only one of two contrasting genetic characteristics
4. The base in the DNA molecule which always pairs off with guanine
5. The shape of a DNA molecule
6. The thread-like structures composed mainly of DNA and proteins and which carry the genes with hereditary factors
7. The synthesis of a mRNA molecule on the pattern of one of the strands of the DNA helix
8. The free triplet of nucleotides on the loop of a tRNA molecule
9. The nitrogenous base which is formed in RNA in the place of thymine
10. The three adjacent nucleotides on a mRNA molecule where transcription occurs during protein synthesis
11. Bead-like proteins in chromosomes attached to DNA
12. Spherical, RNA-containing body within the nucleus of most cells
13. The type of RNA which makes up most of the substances of ribosomes
14. A change in the chemical structure of a gene
15. The process in which DNA molecules make exact copies during the cell cycle
16. The process during which the diploid number of chromosomes is changed to the haploid number
17. The stage in the cell cycle during which the cell performs its normal function
18. The phase in mitosis where the centromere of each chromosome divides into two parts
19. Small RNA molecules in the cytoplasm of cells which collects amino acids to deliver to the ribosomes during protein synthesis
20. The exact copying of the information stored as codons in DNA onto a messenger RNA molecule
21. The complete set of chromosomes found in each nucleus of any individual of a particular species

C. Write down the letter of the description in column B which best suits the term in column A.

Column A

1. Pyrimidines
2. Anticodon
3. Gene mutation
4. Crossing-over
5. A = T; G = C
6. tRNA
7. Transcription
8. mRNA
9. Codon
10. Purines

Column B

A   Produced in nucleus by transcription
B   Constancy in base pairing
C   Complementary base triplets on mRNA strand
D   Cytosine, thymine and uracil
E   Transfers of genetic information from DNA to mRNA
F   Rearrangement in sequence of nucleotides
G   Disrupts gene linkage during meiosis
H   Collects and delivers amino acids to ribosomes
I   Adenine and guanine
J   Base triplets on tRNA molecule which can pair with specific mRNA codon

D. The following six statements refer to the relationship between biological tools and the discoveries made with them. Assume that the tools named in the key, are listed in the order in which they came into use.

For each discovery in the statement, select the earliest tool that made the discovery possible.

KEY: A light microscope   C chemical analysis
     B biological dyes   D electron microscope

1. Within each living cell is a prominent structure, the nucleus
2. The nuclear membrane is composed of proteins and lipids
3. Chromatids are formed during mitosis
4. The nuclear membrane is made of two cell membranes which has many small pores all over its surface
5. The nucleus contains thread-like structures called chromosomes
6. The nucleolus contains RNA

Section B

1. Figure A represents a portion of the uncoiled helix of a molecule. As a result of a certain process, figure B is formed.

    (a) Identify the molecule.
    (b) What is the process referred above, called?
    (c) Where in the cell does this process take place?
    (d) What is the purpose of this process?
    (e) Identify the parts numbered 1 to 6.
    (f) What are the parts numbered 1 to 3 collectively called?
    (g) What is Figure A, on which Figure B is formed, called?

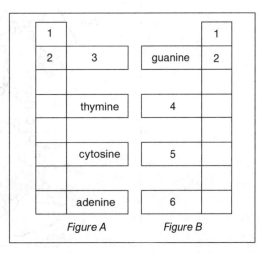

2. The following figure represents a portion of a DNA molecule.
   (a) Identify the parts numbered 1 and 2.
   (b) What are the bonds called which are holding the two DNA strands together?
   (c) What are the monomers of a DNA molecule called?
   (d) Make a labelled diagram of a portion of the mRNA molecule which can be transcribed on this figure as the template.

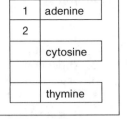

1	adenine
2	
	cytosine
	thymine

3. Tabulate the differences between DNA and RNA.

4. The following diagram illustrates the replication of DNA. Study the diagram and answer the questions.

   (a) During which phase in the cell cycle does this process take place?
   (b) Name **two** substances controlling this process.
   (c) Identify the parts numbered 1 to 6.
   (d) Name the chemical bond at X. What causes this bond to break for unzipping the molecule?
   (e) What is meant by the term "semi-conservative replication"?

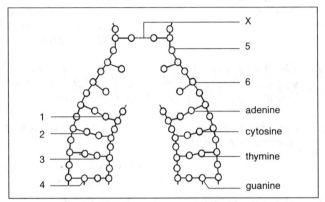

5. The following diagram illustrates the formation of a mRNA molecule using a single strand of a DNA molecule as template. Study the diagram and answer the questions.

   (a) Identify the parts numbered 1 to 12.
   (b) Name the chemical bond at X.
   (c) Where in a cell does the process, illustrated by the diagram, occur and when?
   (d) Name **five** differences between DNA and RNA.
   (e) How do mRNA and tRNA differ with respect to structure and function?

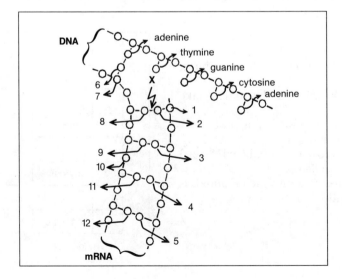

168

6. Study the following diagrams which represent mitosis.

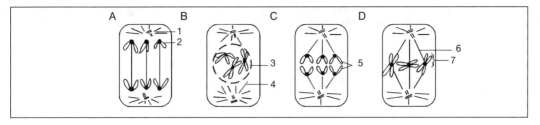

   (a) Identify the phases A, B, C and D.
   (b) Identify the parts numbered 1 to 7.
   (c) What is the end product of this type of cell division?
       How many chromosomes will each new cell have?
   (d) With reference to DNA, describe the events in the nucleus during the interphase of mitosis.
   (e) What happens during cytokinesis?

7. The following diagrams represent three phases in the process of mitosis.

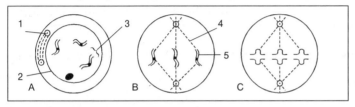

   (a) Identify the phases A to C.
   (b) Identify the parts numbered 1 to 5.
   (c) In which part of the cell is structure No. 1 situated?
   (d) What structures are formed by No. 1 during mitosis?
   (e) Name the gel-like material surrounded by structure No. 2.
   (f) What is the function of No. 3 in the cell?
   (g) With reference to part No. 3 name **three** defferences between mitosis and meiosis I.

8. Study the diagrams below and answer the following questions.
   (a) What type of cell division is illustrated here?
       Give **one** observable reason for your answer.
   (b) Which stage of cell division is illustrated here?
       Give **one** reason for your answer.
   (c) What is the significance of this type of cell division?
   (d) Draw a diagram of one of the resulting daughter cells. Indicate the number and arrangement of chromosomes.

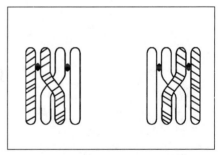

9. Study this diagram showing a section of a type of plant tissue. In your answer use numbers and names.
   (a) Name **one** place where you would expect to find this tissue.
   (b) What process is taking place in the tissue drawn? What use is this process to the plant as a whole?
   (c) Write down the various noticeable stages/phases in the process mentioned in (b), in the **correct; sequence** in which they occur **together** with **one feature** according to which you have recognised each of the stages/phases.

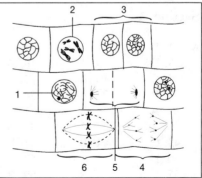

169

10. The following questions are applicable to mitosis and meiosis.
    In a certain plant the somatic cells are diploid and possess **20** chromosomes each. How many of the **following structures** in **each cell** will be **visible** during the phases of mitosis or meiosis as indicated below?
    (a) centromeres during anaphase of mitosis
    (b) centromeres during anaphase I
    (c) daughter chromosomes during anaphase of mitosis
    (d) chromatids during anaphase of mitosis
    (e) chromatids during metaphase I
    (f) chromosomes during metaphase I
    (g) chromosomes at the end of telophase I
    (h) chromosomes at the end of telophase II
    (i) chromosomes during interphase of mitosis
    (j) chromatids during interphase of mitosis

Section C

1. The nucleic acids are amongst the most important substances occuring in living organisms. Discuss their structure, occurence, replication and significance in living organisms.

2. Explain protein synthesis by using a short section of a DNA molecule with base triplets arranged as follows:
   ATC   CGT   GGA   CAG

3. The process of cell replication involves the following main events:
   (a) the formation of the spindle;
   (b) the behaviour of the chromosomes;
   (c) the division of the cytoplasm; and
   (d) the events during the interphase.
   Discuss these events and state their biological significance.

HIGHER GRADE
ANSWERS

Section A

A. 1. D   2. A   3. C   4. C   5. D   6. B   7. B
    8. D   9. A   10. A   11. C   12. D   13. B   14. D
    15. C   16. B   17. A   18. C   19. A   20. B   21. D
    22. B   23. A   24. C   25. B   26. C   27. D   28. C
    29. A   30. B   31. C   32. C   33. D   34. B   35. B
    36. A   37. A   38. B   39. C   40. A   41. B   42. D
    43. C   44. B   45. D   46. D   47. B   48. D   49. B
    50. D   51. C   52. A   53. D   54. D   55. A   56. C
    57. C   58. B   59. B   60. C

B.
1. Nucleotide
2. Adenine
3. Meiosis
4. Cytosine
5. Double helix
6. Chromosomes
7. Transcription
8. Anticodon
9. Uracil
10. Codon
11. Histone
12. Nucleolus
13. rRNA
14. Gene mutation
15. Replication
16. Meiosis
17. Interphase
18. Anaphase
19. tRNA
20. Transcription
21. Genome

C. 1. D 2. J 3. F 4. G 5. B
6. H 7. E 8. A 9. C 10. I

D. 1. A 2. C 3. B 4. D 5. B 6. C

Section B

1. (a) DNA (b) DNA replication (c) Nucleus
   (d) • For daughter cell to have identical DNA composition as that of mother cell after mitosis
   • Ensures that hereditary information in original chromosome of mother cell is passed on to daughter cells when mitosis occurs
   (e) 1 – deoxyribose   2 – phosphate ion   3 – cytosine
   4 – adenine   5 – guanine   6 – thymine
   (f) Nucleotide   (g) Template

2. (a) 1 – deoxyribose   2 – phosphate ion
   (b) Hydrogen bonds
   (c) Nucleotides
   (d)

uracil	1	— ribose
	2	— phosphate ion
guanine		
adenine		

3.

DNA	RNA
1. Double-stranded polymer of nucleotides	1. Single-stranded polymer of nucleotides
2. Large molecular mass and molecule is long	2. Small molecular mass and molecule is short
3. Always a double helix	3. Single or double helix
4. Pentose sugar deoxyribose	4. Pentose sugar is ribose
5. Nitrogenous base is thymine	5. Nitrogenous base is uracil
6. A to T; and G to C = 1:1	6. Bases occur in any number; it varies
7. Occurs in nucleus only	7. Produced in nucleus but found throughout the cell
8. One basic form only, but almost inifinite variety within that form	8. Three basic forms: mRNA, tRNA and rRNA
9. Permanent or long-lived	9. Mainly temporary; short-lived
10. Base-paring rule exists	10. No base-pairing occurs

4. (a) Interphase (b) Proteins and enzymes (DNA polymerases)
   (c) 1 – adenine   2 – guanine   3 – thymine
       4 – cytocine   5 – phosphate ion   6 – deoxyribose
   (d) Hydrogen bond • Enzyme action
   (e) • Parent strand is retained or conserved
       • this conserved strand acts as template
       • new strand is complementary to old strand
       • each 'new' DNA molecule is really half-old, half-new.

5. (a) 1 – ribose   2 – uracil   3 – guanine   4 – cytosine   5 – adenine   6 – thymine
       7 – deoxyribose 8 – adenine 9 – cytosine 10 – phosphate ion 11 – guanine 12 – thymine
   (b) Weak hydrogen bond
   (c) Chromosomes in nucleus; during interphase
   (d) Refer to Question 3
   (e) **mRNA**: Long single-stranded molecule consisting of many nucleotides; carries genetic message or code from DNA to ribosomes; molecule is comparatively short-lived.
       tRNA: Short single-stranded molecule;
       one active codon (anticodon) only with one end folded back on itself;
       transfers amino acids from cytoplasm to ribosomes;
       finds its correct codon for amino acid to be fixed on to mRNA;
       molecule is comparatively long-lived.

6. (a) A – telophase   B – prophase   C – anaphase   D – metaphase
   (b) 1 – centriole   2 – chromatids group themselves at pole
       3 – nuclear membrane breaks down   4 – spindle fibres
       5 – chromatids migrate to opposite poles
       6 – spindle   7 – chromosomes group themselves at spindle equator
   (c) Two daughter cell; three
   (d) See later
   (e) In animal cells invagination of cell membrane occurs; it deepens and cytoplasm splits in two
       in plant cells cell plate forms; middle lamella is formed at equator; cellulose is laid on each side of middle lamella; to complete cell wall and give rise to two separate cells

7. (a) A – prophase B – metaphase C – anaphase
   (b) 1 – centriole 2 – nuclear membrane 3 – chromosome
       4 – spindle fibres 5 – chromatid (c) cytoplasm
   (d) Poles of spindle/microtubules of mitotic spindle (e) Nucleoplasm
   (f) Carriers of genetic information; plays role in protein synthesis
   (g) **Mitosis**: 1. Single chromosomes arrange   **Meiosis**: 1. homologous pairs/bivalents
               on spindle equator   arrange on spindle equator
           2. segments are not exchanged/   2. crossing-over does take place
              crossing-over does not occur
           3. centromeres divide into two   3. centromeres do not divide

8. (a) Meiosis – crossing-over at chiasma occurs
   (b) Prophase I – chromosomes form bivalents from homologous
              pairs   (d)
   (c) • Number of chromosomes is halved to balance doubling effect of fertilisation
       • Genetic variability is increased due to crossing-over
       • produced daughter cells, differentiate into gametes

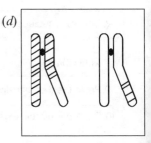

9. (a) Root/stem tip   (b) Mitosis – for growth process
   (c) Prophase   – 2: chromosomes visible and consist of chromatids
       Metaphase   – 6: chromosomes arrange themselves on spindle equator
       Early anaphase – 4: chromatids migrate to opposite poles
       Late anaphase – 5: chromatids arranged at poles
       Telophase   – 3: crosswall develops and cytokinesis occurs
       Interphase   – 1: replication of DNA takes place

10. (a) 40   (b) 20   (c) 40   (d) none – that is already daughter chromosomes
    (e) 40   (f) 20   (g) 10   (h) 10   (i) None   (j) None

Section C

1. *Nucleic acids* – monomers are nucleotides
   Consist of – pentose sugar; phosphate ion; nitrogenous bases

   Occurence and structure

   (a) **DNA**
   - In the genes on chromosomes in nucleus
   - sugar is deoxyribose
   - nitrogenous bases are purines, viz. adenine and guanine
   - and the pyrimidines, viz. cytosine and thymine
   - nitrogenous bases are held together by hydrogen bonds
   - short pyrimidine base combines with long purine base
   - chain of DNA molecule is coiled in double helix
   - helices of DNA molecules are coiled up around histones

   (b) **RNA**
   - Occurs in cytoplasm and nucleolus of the nucleus
   - and on the ribosomes
   - sugar is ribose
   - has uracil instead of thymine
   - is a single chain of nucleotides but chain is much shorter than DNA
   - types of RNA are mRNA, rRNA, and tRNA
   - which are synthesised in the nucleus according to instruction of DNA

   DNA replication
   - During interphase of mitosis
   - DNA helix unwinds gradually
   - by enzyme action weak hydrogen bonds break
   - two strands unzip gradually
   - strands act as template for synthesis of complementary strand of DNA
   - free complementary nucleotides from nucleoplasm
   - become attached to make two identical double strands
   - adenine nucleotide joins with thymine and guanine with cytosine
   - the two double strands each coils up to give
   - two identical DNA molecules identical to original
   - to prevent unwanted replication the DNA molecule assumes a circular form after replication

   Significance
   DNA
   - Holds the hereditary information
     has ability to make exact copies of itself;
     is unchanging and produces the same protoplasm/enzymes;
     in each living cell of given individual

- Controls processes/working of the cell
  like protein synthesis and hence the structure of cell;
  serves as template to synthesise required protein;
  enzymes and several hormones are proteins.

RNA
- Facilitates protein synthesis at ribosomes
  mRNA transmit instruction for synthesis form DNA to ribosome;
  tRNA carries amino acid from cytoplasm to ribosomes;
  rRNA assists to link mRNA to ribosomes.

2. *DNA provides master plan in code form*
   - DNA of chromosomes in nucleus provides master plan
   - of the sequence and type of amino acids to be joined
   - double helix unwinds, hydrogen bonds break and strands move apart
   - one of DNA strands serves as template in synthesis of particular protein
   - base triplets/nucleotide triplets code for a particular protein
   - base triplets **ATC, CGT, GGA** and **CAG** of the DNA molecule
   - are known as codons and form the base of the genetic code

Messenger RNA is transcribed in nucleus
- One of DNA strands serves as template and DNA is copied
- by using complementary RNA nucleotides
- process known as transcription
- free RNA nucleotides fit onto opposite end of this DNA strand
- as result of their complementary relationship cytosine pairs with guanine and adenine with uracil
- the codons which are formed are **UAG, GCA, CCU** and **GUC**
- once mRNA is transcribed it separates from DNA template
- mRNA which is a copy of a short length of a DNA molecule
- leaves the nucleus through pores in nuclear membrane
- and moves to cytosol of cell

Transfer RNA brings amino acids to ribosomes
- mRNA attaches to surface of ribosomes
- it provides the code according to which amino acids must join
- amino acids from cytosol are collected by transfer RNA
- and transported to ribosomes
- tRNA possesses only one active codon of three nucleotides
- active codon is known as anticodon
- anticodons for named codons are **AUC, CGU, GGA** and **CAG**
- each tRNA can pick up only one particular amino acid
- anticodon determines which amino acid is taken from the pool
- tRNA with amino acid moves to the ribosome
- tRNA with anticodon **CAG**, transport amino acid valine only

Amino acids join at ribosome
- Due to complementary relationship the anticodon of tRNA molecule
- fits only with one specific codon of mRNA on ribosome
- each tRNA molecules moves along the mRNA molecule
- and finds the correct codon on the mRNA
- and releases its amino acid in the correct place onto polypeptide chain
- amino acids join in the correct sequence with one another
- and protein molecule is synthesised by ribosome
- once it has deposited its amino acid tRNA moves off
- to cytosol to find new molecule of same sort amino acid
- mRNA can be used over and again
- it synthesises the same protein until demand is satisfied
- several ribosomes can move simultaneously along mRNA strand
- clusters of ribosomes are known as polyribosomes or polysomes

3. (*a*) *Formation of the spindle*
- During prophase the nuclear membrane and nucleolus disappear
- centrioles divide and move to opposite poles
- a number of microtubules form which
- are responsible for the formation of mitotic spindle
- the fibrils stretch cone-shaped from pole to pole
- some fibrils are continuous and
- others are connected to the sentromeres of chromatids

(*b*) *Behaviour of chromosomes*
- DNA strands of chromatin network coil up and become shorter and thicker
- visible as chromosomes each consisting of two chromatids
- joined to each other by means of a centromere
- during metaphase chromosomes arrange on spindle equator
- centromeres lie on spindle equator
- during anaphase centromere divides into two equal parts
- at this stage chromatids are called daughter chromosomes
- microtubules shorten and chromatids are pulled apart
- and move to opposite poles
- during telophase daughter chromosomes group themselves at poles
- the single stranded chromosomes uncoil
- lenghten and fade back as chromatin network

(*c*) *Division of cytoplasm*
- Known as cytokinesis
- in animal cells invagination occurs in vicinity of equator
- the constriction of cell membrane deepens and
- cytoplasm divides into two each with its own nucleus
- in plant cells a cell plate develops
- middle lamella forms at equator
- cellulose is laid down at each side of middle lamella
- cell wall forms and two separate cells are produced

(*d*) *Interphase*
- This is the period of cellular synthesis and growth
- includes synthesis of RNA, proteins
- and materials for mitotic spindle and DNA replication
- and duplication of chromosomes, centrioles
- as well as organelles, such as ribosomes, mitochondria, plastids
- energy is supplied by cellular respiration

(*e*) *Significance*
- Mitosis is process which allows multicellular organisms grow in size
- ensures exact number of chromosomes and identical DNA composition to daughter chromosomes
- allows for replacement of worn out and damaged tissues
- in most unicellular organisms asexual reproduction is by mitosis
- is responsible for growth and development of structures which reproduce vegetatively, e.g. cuttings

9 Genetic Mechanisms

Genetics is the science of **inheritance**, the way in which structural and functional characteristics are transmitted from parent to offspring. In living plant and animal cells, **hereditary instructions** are encoded in DNA molecules.

These instructions have two striking properties:
(a) they assure that the offspring will **resemble** their parents (due to mitotic nuclear division and asexual reproduction); and
(b) they allow scope for **genetic variations** in the detail of their **characteristics** or traits (due to meiotic nuclear division and sexual reproduction).

DNA holds the instruction for producing and controlling all the living body cells.

1. *Gametes as vehicles of inheritance*

    Gametes are the vehicles of genetic information and inheritance, and each has one complete set of DNA molecules in its **chromosomes**. The zygote which is produced when the male and female gametes fuse, will therefore be provided with a **double set** of chromosomes and will be diploid. The new individual has two sets of chromosomes – one set from its father and one set from its mother.

    The single long **DNA molecule** in each chromosome comprises numerous shorter sections, called **genes**. The gene is the basic unit of inheritance and consists of a large number (as many as 3 000) **nucleotides**. Genes are **specific segments** of the DNA strand and they contain the hereditary instructions for producing or influencing a trait in the offspring. Both parents pass on one of each gene to the offspring by means of meiosis, gamete production and fertilisation. Thus the first cell (zygote) of a new individual inherits **two genes for every trait**, that is one from each parent, one on each of the two homologous chromosomes. **Homologous chromosomes** have the same length, same centromere location and the same genes. They line up with each other during meiosis.

    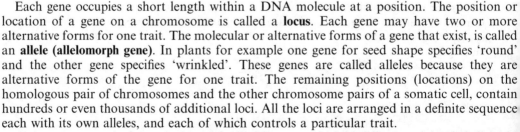

    A chromosome.

    Although both genes deal with the same trait, they may differ in their information about it, because of slight molecular differences. As a result of **meiosis** only one of the pair of genes can be in any one gamete.

    Each gene occupies a short length within a DNA molecule at a position. The position or location of a gene on a chromosome is called a **locus**. Each gene may have two or more alternative forms for one trait. The molecular or alternative forms of a gene that exist, is called an **allele (allelomorph gene)**. In plants for example one gene for seed shape specifies 'round' and the other gene specifies 'wrinkled'. These genes are called alleles because they are alternative forms of the gene for one trait. The remaining positions (locations) on the homologous pair of chromosomes and the other chromosome pairs of a somatic cell, contain hundreds or even thousands of additional loci. All the loci are arranged in a definite sequence each with its own alleles, and each of which controls a particular trait.

    Gene assortment during meiosis and fertilisation may result in different mixings of alleles in the offspring. When two alleles of the gene pair at the same locus on homologous chromosomes are identical, they are said to be **homozygous** (pure-breeding). If different, it is said to be **heterozygous** (hybrid). An individual which is heterozygous for a gene, normally displays the trait determined by one of the allelic pair. This is called the **dominant** allele and its partner which does not display the trait is called the **recessive** allele. Capital letters is used for dominant alleles and lower-case letters for recessive ones, e.g. alleles A and *a*; B and *b*. A **homozygous dominant** individual has two dominant alleles (AA) for that trait and a **homozygous recessive** individual has two recessive alleles (*aa*). A **heterozygous** individual has two different alleles (A*a*).

The genetic composition of an organism is called the **genotype**, that is the **genes** present in an individual (AA, *aa*, A*a*). The **phenotype** is an organism's external appearance, that is an individual's observable traits. The phenotype of an individual is determined by its genotype.

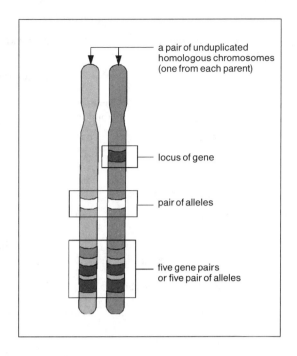

2. *Monohybrid crosses*

Mendel used in his experiments varieties of the garden pea that were pure-breeding. In his first investigations he studied the inheritance of a single pair of contrasting characteristic (trait) – **monohybrid crosses**. This is known as his first law of inheritance or the **Law of Segregation**. When gametes are formed alleles separate or segregate. He showed that the F_1 generation always had the **dominant** allele. In the F_2 generation 75% of the offspring will have the **dominant** characteristic and 25% the **recessive** characteristic.

In his first investigation Mendel studied the inheritance of a single pair of contrasting characteristics, that is **monohybrid inheritance**. From it can be con-ducted that:

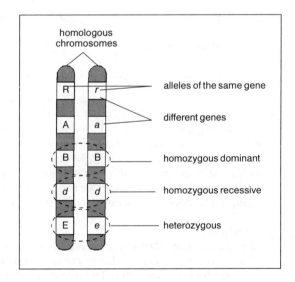

(i) the genes occur in pairs and may be dominant or recessive with respect to one another; and
(ii) only one such gene may be carried in a single gamete.

Genetic representation of a monohybrid cross

R = allele for **round** seed; *r* = allele for **wrinkled** seed

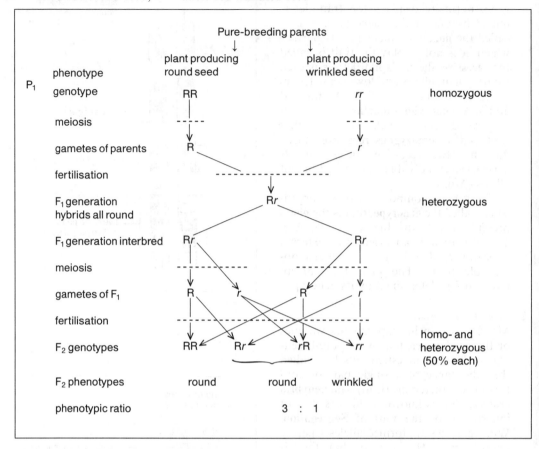

3. *Dihybrid crosses* (HG only)
   In later experiments Mendel studied the inheritance of two pairs of characteristics, that is **dihybrid inheritance**. In dihybrid crosses he concluded that each member of one pair of contrasting characteristics segregates independently of that of any other pair. This is embodied in Mendel's second law, the **Law of Independent Assortment**. This law applies only to separated traits which are on different chromosomes. On the same chromosome they form **linkage groups**. The conclusion drawn from the dihybrid cross can be interpreted in terms of **meiosis**.

4. *Dominance*
   Crosses between **homozygous** plants and animals which differ in a pair of **contrasting characteristics**, produce a **heterozygous generation** showing only one trait, the **dominant**, and not the other, the **recessive**. The dominant characteristic masks the expression of the recessive one. In garden peas round seed, yellow seed and tall plants are dominant and wrinkled seed, green seed and short plants are recessive.

   Incomplete dominance also occur between alleles and hybrids show a mixture of these traits. When a **homozygous dominant** red-flowered (RR) petunia (or snapdragon) is crossed with a **homozygous recessive** white-flowered (*rr*) petunia, all the F_1 flowers are pink (R*r*). The phenotype of the pink heterozygote is **intermediate** between red and white. This indicates a lack of dominance for both red and white alleles. Subsequent crossing of the F_1 offspring produces red, pink and white flowers in a monohybrid genotypic ration of **1:2:1**. The dominant allele requires a red pigment, and it takes two alleles to produce enough pigment to

give the flower a red colour. With their single dominant allele, heterozygotes can only produce enough pigment to give the flower a pinkish colour. Another example of incomplete dominance is the ABO blood group system in the human where an AB blood group is possible.

5. *Segregation*
Contrasting characteristics are determined by sets of alleles. Diploid organisms inherit a **pair of alleles** for each trait. The pair of alleles is located on **homologous chromosomes**. During **meiosis** the two alleles **segregate** from each other, so each gamete formed after meiosis, has an equal chance of receiving one or the other gene, but not both. Therefore, each gamete contains **only one gene** for a particular characteristic.

6. *Independent assortment* (HG only)
According to the **Law of Independent Assortment** each member of an allelic pair may combine randomly with either of another pair. This is also known as **dihybrid inheritance**. Each gene pair tends to **assort gametes independently** of other gene pairs that are located on non-homologous chromosomes. The independent inheritance of various characteristics results from the random positioning of homologous chromosomes during **metaphase I** of meiosis. This leads to the random distribution of alleles during gamete production.

7. *Sex determination*
Homologous chromosomes are normally identical in appearance and are called **autosomes**. An exception is provided by the **sex chromosomes**, known as **gonosomes** with the sex genes. Sex is determined by a pair of alleles carried on the **X** and **Y** gonosomes. In man the female is **XX** and the male **XY**.

8. *Proteins and determination of characteristics* (HG only).
Most of the facts of gene action can be explained on the basics of alleles producing **proteins**. Proteins produced during protein synthesis are used as **fundamental structural** components of cells, chromosomes and enzymes. Therefore, the **nature of the proteins** which can be produced in a cell, will play a determining role in the development of structural and physiological **characteristics** of a cell and of an individual.

9. *Gene mutation* (HG only)
Gene mutations are **changes of the base-sequence** within the DNA molecule and therefore, in the genotype of organisms. Such mutations result in different informational content of DNA which is transmitted through replication to cell progeny or by transcription to mRNA. A mutation may involve a **codon substitution**, the **deletion** of one or more nitrogenous bases from a codon, or the **insertion** of additional bases into the DNA double-helix.

When an allele of a gene is mutated into the new allele, it **tends to be recessive** and its effects are normally masked by its partner allele. Only in the homozygous condition can such mutant genes express their phenotypic effect. A new, mutated DNA chain may differ from the old one by **only a single nucleotide**. This altered gene is called a **mutant allele**, and is replicated along with the normal DNA and may be transmitted to future generations. Examples of gene mutations are sickle-cell anaemia, haemophilia and albinism.

Gene mutation enables individuals of the same species to differ from one another. These differences may ensure that individuals of one species remain adapted to their environment. Their favourable characteristics (genes) will be transmitted to subsequent generations. Natural selection contributes to the survival of the species.

10. *Natural selection* (HG only)
Natural selection depends on variation and is the selection by the enviroment of individuals with the most **beneficial hereditary variations**, both those that already exist and whatever new variations arise. Nature selects those individuals **best fitted for survival** in their environment. This is done by selecting the most **useful characteristics** in a species and allows them to be passed on to subsequent generations.

Variation allows for the gradual increase in the **gene pool** of a population. All the possible alleles for all the genes of a given population represents the total gene pool. There is a gradual increase in this gene pool of those genes which produce favourable characteristics, and a corresponding decrease of alleles that result is less favourable traits.

11. *Practical applications of genetics*
    Man selects those individuals with the most useful characteristics and allowing them to breed. This is **artificial selection** and is used by plant and animal breeders. The result is not a population better adapted to its enviroment, but a population better adapted to our needs. Outbreeding produces tougher individuals with a better chance of survival and which are more productive. This is called **hybrid vigour** or **heterosis**. Mules are produced from a cross between a horse and a donkey.
      Inheritance of **blood groups** and of **Rhesus factor** in humans is genetically predictable. This is very important when blood transfusions are given in cases of illness or accident. **Genetic engineering** uses chemical techniques to affect the DNA molecules in organisms and artificially change their metabolism.

STANDARD GRADE
QUESTIONS

Section A

A. *Various possibilities are suggested as answers to the following question. Indicate the correct answer.*

1. Genetics is
   A   gamete production.
   B   DNA replication.
   C   the synthesis of proteins.
   D   the science of inheritance.

2. After fertilisation the zygote inherits
   A   one gene for every trait.
   B   two traits for every gene.
   C   two genes for every trait.
   D   one trait for every gene.

3. A linear collection of genes is called a
   A   nucleotide.
   B   chromosome.
   C   allele.
   D   chromatid.

4. A gene is the
   A   basic unit of inheritance.
   B   one partner of a homologous chromosome pair.
   C   locus.
   D   alternative molecular form of a chromosome.

5. Genes do **not** occur in pairs in
   A   cells of plant leaves.
   B   gametes.
   C   somatic cells.
   D   the zygote.

6. The position of a gene on a chromosome is called
   A   a locus.
   B   an allele.
   C   genetics.
   D   allelomorph gene.

7. The molecular or alternative forms of genes are called
   A   bivalents.
   B   phenotypes.
   C   alleles.
   D   genotypes.

8. The external appearance of an individual due to its genetic composition.
   A   Homozygous
   B   Genotype
   C   Heterozygous
   D   Phenotype

9. The genetic composition of an organism is called the
   A   homozygote.
   B   genotype.
   C   heterozygote.
   D   phenotype.

10. Gene assortment takes place during
    A   meiosis only.
    B   fertilisation only.
    C   mitosis.
    D   meiosis and fertilisation.

11. A homozygous dominant individual has for a trait
    A   one dominant allele.
    B   one dominant and one recessive allele.
    C   two dominant alleles.
    D   two recessive alleles.

12. A homozygous recessive individual has for a trait
    A   two recessive alleles.
    B   one recessive and one dominant allele.
    C   two dominant alleles.
    D   one recessive allele.

13. A heterozygous individual has for a trait
    A   one dominant allele.
    B   one dominant and one recessive allele.
    C   two dominant alleles.
    D   two recessive alleles.

14. The genotype of a homozygous dominant individual for a particular trait is
    A   A.
    B   Aa.
    C   AA.
    D aa.

15. The genotype of a heterozygous individual for a particular trait is
    A   A.
    B   Aa.
    C   AA.
    D aa.

16. The genotype RR, is an example of a
    A   hybrid.
    B   gamete.
    C   heterozygote.
    D   homozygote.

17. The probability that two parents, heterozygous for an inherited trait will have a homozygous dominant offspring is
    A   1 in 4.
    B   1 in 3.
    C   1 in 2.
    D   1 in 1.

18. In a certain plant Aa is the genotype of a particular characteristic. After **mitosis** has taken place, the genes of the daughter cells will be as follows:
    A   each possesses A only.
    B   each possesses a only.
    C   each cell possesses Aa.
    D   one cell possesses AA and the other one aa.

19. When the allele is different on both homologous chromosomes, the organism is
    A   a homozygote.
    B   a heterozygote.
    C   sterile.
    D   recessive.

20. To illustrate dominance the following should be crossed, viz.
    A   Rr x rr.
    B   Rr x Rr.
    C   RR x Rr.
    D   RR x rr.

21. A pea plant bearing red flowers is crossed with a pea plant with white flowers. The flowers of all the F_2 plants are red. Which of the following represents the genotype of the parents? (R = red; r = white)
   A   Rr x rr
   B   RR x Rr
   C   rr x Rr
   D   Rr x Rr

22. The offspring of a pure breeding brown mouse which had been crossed with a pure breeding white mouse were inbred. Brown is dominant. The expected offspring will be
   A   all brown.
   B   three-quarter white and one quarter brown.
   C   one half brown and the other half white.
   D   three-quarter brown and one quarter white.

23. The alternative gene which occupies the same relative position on a particular chromosome and which controls a particular characteristic, is known as
   A   locus.
   B   allele.
   C   homozygote.
   D   heterozygote.

24. The genetic composition of plants resulting from a cross between a plant with red flowers (RR) and one with white flowers (WW) with no dominance is
   A   Rr.
   B   Rw.
   C   WW.
   D   RW.

25. Which is **not** a possible genotype of the offspring of Rr x Rr?
   A   R
   B   rr
   C   RR
   D   Rr

26. Which of the following will give a 1 : 1 phenotype ratio in the F_1 generation?
   A   Aa x Aa
   B   AA x aa
   C   Aa x aa
   D   AA x Aa

27. In humans a child's sex is normally determined by
   A   the mother's gamete.
   B   the autosomes.
   C   the father's gamete.
   D   the XX chromosomes of the mother.

28. A zygote with XX gonosomes
   A   is haploid.
   B   will give rise to a female individual.
   C   will give rise to a male individual.
   D   is not possible.

B. *Write down the correct term for each of the following statements.*

1. The science of inheritance
2. The father of genetics
3. The offspring of two genetically dissimilar individuals
4. A zygote, having a pair of homologous chromosomes, both with a recessive characteristic
5. A genetic crossing involving only one contrasting character
6. Cells which fuse to form a zygote
7. Any chromosome which is not a sex chromosome
8. The XX and XY sex chromosomes which are responsible for the inheritance of sex
9. The 44 ordinary homologous chromosomes in the body cells of man
10. The external appearance of an organism which is the result of its environmental factors
11. The genetic make-up of an organism
12. The position of a gene on a chromosome
13. The alternative form of a gene occupying the same locus on homologous chromosomes
14. Hybrids having desirable characteristics lacking in both parents

C. **Write down the letter of the description in column B which best suits the term in column A.**

Column A		Column B
1. Heterozygous	A	An organism's external appearance
2. Genotype	B	Having two identical genes for a particular trait
3. Hybrid	C	Genes that an organism contains
4. Homozygous	D	Having two different genes for a particular trait
5. Phenotype	E	Cross between two genetically different parents

D. **The gene for brown coat colour (B) is dominant over that for white coat colour (b) in rats. Two brown coated rats were mated.**

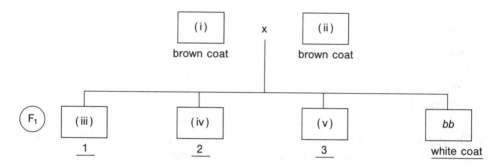

(a) Write down the genotypes of the parents and of the F_1 generation numbered (i) to (v).
(b) Write down the rest of the phenotypes of the offspring in the F_1 generation numbered 1 to 3.
(c) What percentage of the F_1 generation is heterozygous for brown coat colour?

Section B

1. In man the gene for brown eyes is dominant to the gene for blue eyes. A blue-eyed man marries a brown-eyed woman. They have two children.
   (a) Write down genotypes of the man **and** the two children numbered (i) to (iii).
   (b) Write down the phenotype of each of the children numbered (ii) and (iii)
   (c) Is the mother homozygous or heterozygous for brown eyes?
   Give a reason for your answer.

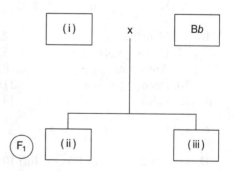

2. A plant homozygous for green leaves (G) is crossed with a plant homozygous for mottled leaves The F_1 generation is green leafed.
   (a) Show the genotypic results of this cross diagrammatically, up to and including the F_2 generation.
   (b) What percentage of the F_2 generation is homozygous for mottled leaves?

3. In guinea-pigs, the gene for black coat is dominant to the gene for white. Two heterozygous black guinea-pigs are crossed.
   (a) By means of a diagram show the genotypic results that would be expected in the F_1 generation.
   (b) One of the white F_1 offspring was crossed with its black parent. By means of a diagram show the expected F_1 genotypic results of this new cross.

4. Study the diagram which shows the results after Mendel had crossed a pure breeding tall garden pea plant with a pure breeding short pea plant. Answer the following questions.

   (a) Let T represents the gene for tallness and t the gene for shortness. Write down the genotypes for each of A, B, E and G.

   (b) Use the same notation to write down the genotype of the F_1 generation.

   (c) What will the phenotype of the generation mentioned in (b) be?

   (d) Which one of these two characteristics would you regard as recessive?

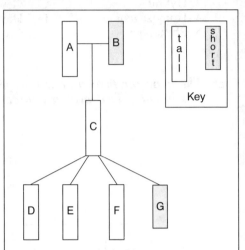

5. (a) Distinguish between the terms autosomes and gonosomes.
   (b) Briefly describe how sex is determined by the sex chromosomes in man.

STANDARD GRADE
ANSWERS

Section A
A. 1. D   2. C   3. B   4. A   5. B   6. A   7. C
    8. D   9. B   10. D   11. C   12. A   13. B   14. C
    15. B   16. D   17. A   18. C   19. B   20. D   21. B
    22. D   23. B   24. D   25. A   26. C   27. C   28. B

B. 1. Genetics   2. Mendel   3. Hybrids
    4. Homozygote   5. Monohybrid   6. Gametes
    7. Autosome   8. Gonosomes   9. Autosomes
    10. Phenotype   11. Genotype   12. Locus
    13. Allele   14. Heterosis/hybrid vigour

C. 1. D   2. C   3. E   4. B   5. A

D. (a) (i) Bb (ii) Bb (iii) BB   (iv) Bb (v) Bb
    (b) 1 – brown   2 – brown   3 – brown   (c) 50%

Section B

1. (a) (i) bb (ii) Bb (iii) bb
    (b) (ii) Brown   (iii) Blue
    (c) Heterozygous – has a gene for brown and one for blue.

2. (*a*) G = dominant gene for green leaves;
g = recessive gene for mottled leaves

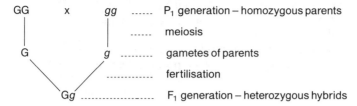

Colour of F_1 generation is green

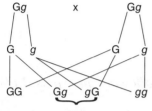

GG	Gg gG	gg
green homo- zygous	green hetero- zygous	mottled homo- zygous

3 : 1 – ratio

(*b*) 25%

3. (*a*) B = black; *b* = white

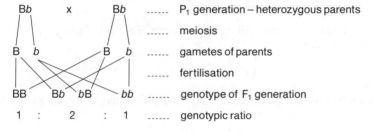

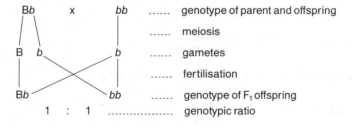

4. (*a*) A = TT B = tt E = T*t* G = tt
(*b*) T*t* (*c*) Tall (*d*) Short

5. (*a*) **autosomes**: chromosomes other than sex chromosomes
gonosomes: the sex chromosomes
(*b*) Somatic cells in male consist of 22 pairs autosomes and XY gonosomes
Somatic cells in female consist of 22 pair autosomes and XX gonosomes

During meiosis homologous chromosomes separate/segregate
So that one half of male gametes will have 22 autosomes and an X gonosome
While other half will have 22 autosomes and a Y gonosome
The female gametes all will have 22 autosomes and an X gonosome
During fertilisation the following is possible:
if 22 + Y and 22 + X fuse, the zygote will develop into a male
OR 22 + X and 22 + X fuse, the zygote will develop into a female

HIGHER GRADE
QUESTIONS

Section A

A. Various possibilities are suggested as answers to the following questions. Indicate the correct answer.

1. The number of genes for seed colour found in the egg cell of a flowering plant is
   A   1.
   B   2.
   C   hunderds.
   D   thousands.

2. Parents who already have two sons and one daughter, next chance of another daughter, theorically is
   A   25%.
   B   33%.
   C   50%.
   D   75%.

3. In a crossing between heterozygous parents, **300** offspring were produced. How many of these were homozygous for the dominant gene?
   A   225
   B   150
   C   100
   D   75

4. The seeds produced when a heterozygous pea plants are cross-pollinated, will yield plants with a genotype ratio of
   A   3 : 1.
   B   9 : 3 : 3 : 1.
   C   1 : 2 : 1.
   D   1 : 1.

5. Hybrid means the same as
   A   mutant.
   B   dominant.
   C   recessive.
   D   heterozygous.

6. A chemical change in a gene that alters its hereditary effects is known as
   A   sex determination.
   B   mutation.
   C   replication.
   D   independent assortment.

7. A pair of chromosomes carries the gene combination Ab and aB respectively **before meiosis**. How can an AB-gamete be explained?
   A   Independent assortment
   B   Crossing over
   C   Segregation of alleles
   D   Chromosome mutation

8. In humans, brown eyes are dominant and blue eyes recessive. What is the percentage possibility that the eyes of the children of parents with heterozygous brown eyes, will be blue?
   A   75%
   B   50%
   C   25%
   D   0%

9. In a cross between two heterozygous plants for red flowers, **80** offspring are produced. How many of the offspring will have white flowers?
   A 80
   B 60
   C 40
   D 20

10. The offspring of a homozygous white guinea-pig which had been crossed with a homozygous black guinea-pig were all grey. What percentage of the F_2 generation will be grey if two grey guinea-pigs were mated?
    A 75%
    B 50%
    C 34%
    D 25%

11. Two cats, heterozygous for the colour white, were crossed. (White hair colour is dominant to black). What would the ratio in the first offspring be as far as hair colour is concerned?
    A All white
    B One half black and the other half white
    C One quarter black and three quaters white
    D One third black, one third white and one third grey

12. A person carries the gene pairs A and A, B and b, and D and d on different chromosome pairs. One of his/her gametes would contain
    A a b d.
    B A B D d
    C A b D.
    D AA Bb Dd.

13. How many heterozygous offspring would you expect if two parents, who were heterozygous for a trait, produced an F_1 generation of **40** individuals?
    A 10
    B 15
    C 20
    D 40

14. Which cross will produce a **9 : 3 : 3 : 1** ratio of phenotypes?
    A aa bb x Aa Bb
    B Aa BB x Aa Bb
    C aa BB x AA bb
    D Aa Bb x Aa Bb

15. A homozygous red-flowered snapdragon is crossed with a homozygous white-flowered snapdragon. The F_1 generation, which produce pink flowers only, is inbred and **20** of the F_1 offspring have white flowers. What will the number and phenotype of the remainder of the F_1 plants be?
    A 10 plants with red flowers
    B 10 plants with pink flowers
    C 20 plants with red flowers and 20 with pink flowers
    D 20 plants with red flowers and 40 with pink flowers

6. If black coat is dominant to white coat, the genotype of which of the following individuals can be determined just by looking at them?
   A Individuals with white coats
   B Individuals that have dominant characteristic
   C Individuals with brown coats
   D Individuals with black coats

7. A heterozygote is
   A a haploid condition in genetic terms.
   B a condition in which both alleles of a pair are different.
   C one of at least two forms of a gene.
   D a condition in which both alleles of a pair are the same.

8. Two genes of a pair on homologous chromosomes end up in separate
   A non-homologous chromosomes.
   B heterozygous chromosomes.
   C body cells.
   D gametes.

B. Write down the correct term for each of the following statements.

1. A zygote having a pair of identical chromosomes, both with either a recessive or a dominant characteristic
2. The alternative form of a gene which is situated on the same locus of homologous chromosomes
3. A genotype consisting of two identical genes for a specific characteristic
4. A sex chromosome that carries information about the production of the primary sexual characteristics of the species
5. All the genes contained within a specified population
6. A change in the chemical structure of a gene
7. A condition of an individual having two different alleles in the corresponding position of homologous chromosomes so that the genotype contains both alleles
8. The offspring of two genetically dissimilar individuals
9. Describes a trait that will not appear in the phenotype when present in the heterozygous condition
10. A genetic crossing involving two contrasting characterisistics

C. Write down the letter of the description in column B which best suits the genotypes in column A.

Column A		Column B
1. Aa x Aa	A	True-breeding parents
2. Aa Bb x Aa Bb	B	Homozygous condition
3. Aa	C	Dihybrid cross
4. AA x aa	D	Heterozygous condition
5. aa	E	Monohybrid cross

Section B

1. The diagram below shows the pedigree of two rabbit families. The allele for white fur (b) is recessive. Study the diagram and answer the questions.
   KEY: squares = males   shaded symbols = black fur
           circles = females   unshaded symbols = white fur

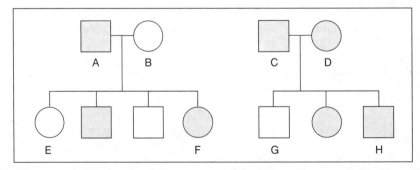

   (a) State the genotype of each rabbit numbered A to H.
   (b) If rabbit F and rabbit G were crossed, what percentage of the offspring would have:
      (i) black fur, and
      (ii) white fur?

2. In humans, a gene responsible for clotting blood is carried on the X-chromosome. People who carry only the recessive genes bleed easily and are called hemophiliacs. The diagram below shows the occurrence of hemophiliacs in a certain family.

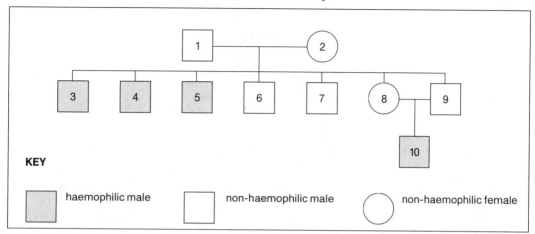

Indicate and explain the genotype for each individual numbered in the diagram.

3. Red colour (RR) in a certain flower is dominant to white (rr). Two homozygous flowers, one red and one white, are crossed.
   (a) What is the phenotype and genotype of the F_1 generation?
   (b) Make a labelled diagram to illustrate the phenotype and genotype of the F_2 generation if the F_1 generation is allowed to self-pollinate.
   (c) State Mendel's Law of Segregation proven in this experiment.

4. In horses black coat colour (B) is dominant to white (b). A white mare mates twice with the same black stallion. She produces a white foal on the first occasion and a black foal on the second occasion.
   Use the letters B and b as indicated above and write down the genotypes of:
   (a) the mare;
   (b) the stallion;
   (c) the first foal; and
   (d) the second foal.

5. The diagram below shows the inheritance of eye colour in humans. The squares represent men and the circles, women. The individuals represented by shaded symbols have brown eyes, and the unshaded symbols blue eyes.
   Brown eye colour (B) is dominant to blue (b).

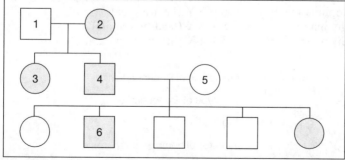

   (a) Use the letters B and b, as indicated, and write down the genotypes of the individuals numbered 1 to 5.
   (b) Draw a diagrammatic representation of all the genetic combinations in regard to eye colour, of the descendants when No. 6 marries a woman with the same genetic composition as No. 3. Use the letters B and b to show the genetic composition.

6. A tall sweet pea plant with yellow seeds is crossed with a dwarf sweet pea plant with green seeds. Tall (T) and yellow (G) are dominant and dwarf (*t*) and green (*g*) are recessive traits.
   (*a*) Use the information above to determine the genotypes of the F_2 generation.
   (*b*) What will the ratio of the various phenotypes in the F_2 generation be?

HIGHER GRADE
ANSWERS

Section A

A. 1. A 2. C 3. D 4. C 5. D 6. B 7. B
    8. C 9. D 10. B 11. C 12. C 13. C 14. D
    15. D 16. A 17. B 18. D

B. 1. Homozygote 2. Allele 3. Homozygous
    4. Gonosome 5. Gene pool 6. Gene mutation
    7. Heterozygous 8. Hybrid 9. Recessive
    10. Dihybrid crossing

C. 1. E 2. C 3. D 4. A 5. B

Section B

1. (*a*) A – B*b* B – *bb* C – B*b* D – B*b* E – *bb*
          F – B*b* G – *bb* H – BB/B*b*
   (*b*) (i) 50% (ii) 50%

2. Haemophilia
   • carried by a recessive allele occuring only on the X chromosome (indicated as **X**)
   • female can only be a haemophiliac if double recessive (**XX**)

   X – recessive gene containing the defective recessive allele
   X – dominant gene opposing the haemophilic characteristic
   Y – contains no dominant allele

   1 – XY (healthy male)   2 – XX (haemophilic female)
   3 – XY (haemophilic male)   4 – XY (haemophilic male)
   5 – XY (haemophilic male)   6 – XY (healthy male)
   7 – XY (healthy male)   8 – XX (healthy female)
   9 – XY (healthy male)   10 – YX (haemophilic male)

3. (*a*)

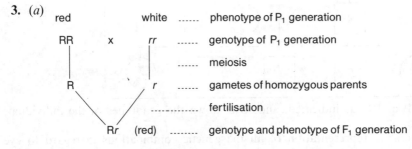

    red   white   ...... phenotype of P_1 generation
    RR   x   rr   ...... genotype of P_1 generation
                            ...... meiosis
    R   r   ...... gametes of homozygous parents
                            ...... fertilisation
         R*r* (red)   ...... genotype and phenotype of F_1 generation

    phenotype – red; genotype – R*r*

(b) 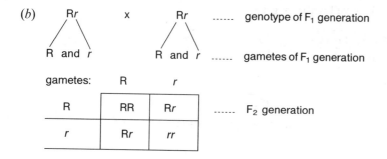

genotype = 1 RR : 2 Rr : 1 rr;   phenotype = 3 red : 1 white

(c) The hereditary factors (genes); become separated; when sex cells are produced; gamete possesses only one of two contrasting characteristics

4.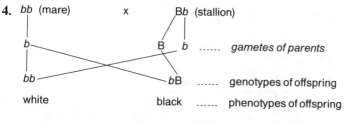

(a) bb (b) Bb (c) bb (d) Bb

5. (a) BB = brown eyes; bb = blue eyes
since both No. 3 and No. 4 have brown eyes, No. 2 will be **BB**

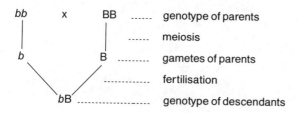

1 = bb 2 = BB 3 = Bb 4 = Bb 5 = bb

(b)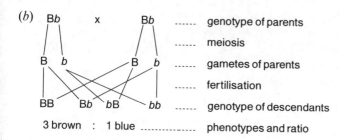

3 brown : 1 blue .... phenotypes and ratio

6. (a) T = tall; t = dwarf; G = yellow; g = green

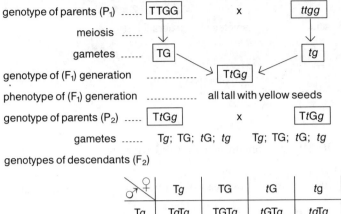

♂ \ ♀	Tg	TG	tG	tg
Tg	TgTg	TGTg	tGTg	tgTg
TG	TgTG	TGTG	tGTG	tgTG
tG	TgtG	TGtG	tGtG	tgtG
tg	Tgtg	TGtg	tGtg	tgtg

(b) Phenotypes of descendants (F_2)
9 tall yellow; 3 tall green; 3 dwarf yellow; 1 dwarf green

10 Reproduction in Humans

If a species was to stop reproducing, it would eventually become **extinct**. Species survive by reproduction. **Sexual reproduction** in humans is a complex biological process. It is further complicated by **personal** and **social factors**. Human beings do not have a special reproduction season. Biologically, human beings can be sexually active from early youth until a very high age. Sexuality in humans is much more than producing offspring; it is a kind of bond that keeps communities together. Sexuality influences what we think of other people and how we act towards other people.

In humans, the ability to reproduce starts with **puberty**. During this time, changes take place in the human body making it physically possible to reproduce.

1. Reproductive organs

(a) *Male reproductive organs*

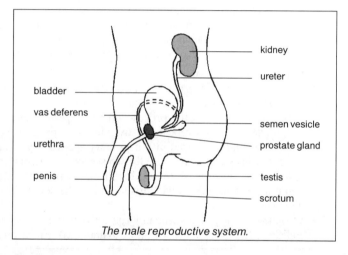

The male reproductive system.

The primary male sex organs comprise two **testes** situated in a bag, the **scrotum**, between the legs. The testes contain the seminiferous tubules (vasa efferentia). The seminiferous tubules are lined with **germinal epithelium** which by meiosis gives rise to **sperms**. In the germinal epithelium are the **cells of Sertoli** which supply the developing sperms with nutrition and the **cells of Leydig** which produce the male sex hormone, **testosterone**. Each sperm has a long tail, with which it swims in the **seminal fluid**, and a **head** with the **nucleus** and **acrosome**.

The sperms are temporarily stored in the **epididymis**. The tubes of the epididymis join to form a single tube, the **sperm ducts** or **vas deferens** which runs through the abdominal cavity and eventually opens into the **urethra**. The urethra opens at the tip of the **penis**, the male copulation organ, through which the sperms are introduced into the **vagina** of the female.

The **prostate gland** lies where the vas deferens joins the urethra. The **seminal vesicle** also joins the urethra more or less at the same point. The secretions of these two glands contain nutrients and enzymes which stimulate the sperms to greater motility. An **alkaline fluid**, secreted by the prostate gland, protects the sperm against acids of the female vagina. The sperms and above-mentioned substances are collectively known as **semen**.

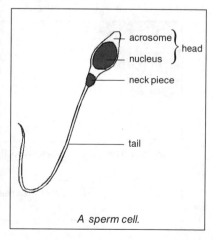

A sperm cell.

193

(b) *Female reproductive organs*

The primary **female reproductive organs** consist of two **ovaries** situated in the lower part of the abdominal cavity just below the kidneys, on either side of the **uterus**. Near each ovary is a funnel-shaped opening of the **Fallopian tubes** leading to the uterus. Each ovary is surrounded by a layer of **germinal epithelium** that divides during the embryonic stage to produce thousands of units, the **Graafian follicles**. After **ovulation**, the remains of the Graafian follicles are converted into a mass of cells, the **corpus luteum**, which produces the female hormone, **progesterone**.

The female reproductive system.

The egg cell is collected by the funnel of the **Fallopian tube** and sucked into the funnel by the **fimbriae**. The two Fallopian tubes, one on either side, lead to the **uterus** where the embryo can develop further. The uterus narrows to a glandular **cervix** that leads to a slightly wider tube, the **vagina**, that can accommodate the **penis** during sexual intercourse. The vagina opens to the exterior through the **vulva**. The vulva is a collective name for the external genital organs, which consist of two outer folds of skin, the **labia majora**, covering two inner more delicate folds, the **labia minora**. In a virgin the opening of the vagina is partly closed by the **hymen**.

2. **Gametogenesis** (HG only)

Gametogenesis is the process whereby **adult gametes** are produced in the **gonads**.

(a) *Spermatogenesis*

Spermatogenesis is the process whereby **sperm cells** are produced in the **testes**. The germinal epithelium divides repeatedly by **mitosis** to produce **spermatogonia** (2n). A spermatogonium enlarges and develops into a **primary spermatocyte** (2n). Each primary spermatocyte undergoes **meiosis I** and gives rise to two **secondary spermatocytes** (n). Each secondary spermatocyte undergoes **meiosis II** to produce four **spermatids** (n). Each spermatid differentiates into a mature **sperm cell** (n) which is stored in the seminiferous tubules and epididymis.

(b) *Oogenesis*

Oogenesis is the process whereby **egg cells** (ova) are produced in the **ovaries**. At birth each **primary follicle** in the ovary contains an **oogonium** (2n). From puberty the oogoniums grow and develop

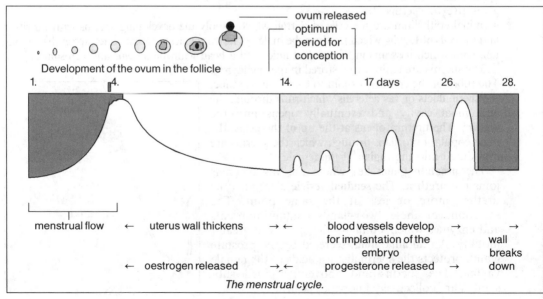

The menstrual cycle.

into **primary oocytes** (2n). The primary oocytes only form under the influence of the hormone, **FSH**, of the **hypophysis**. The ripe follicle is known as a **Graafian follicle**. It moves to the outer wall of the ovary, breaks open and **meiosis I** takes place. The products of the first meiotic division are a **secondary oocyte** (n) and a **polar body** that are released every 28 days by **ovulation**. Ovulation in the human takes place almost halfway through this cycle, known as the **menstrual cycle**. The moment when a sperm penetrates a **secondary oocyte** after sexual intercourse, the **second meiotic** division takes place to form an **ovum** and a **second polar body**. Fertilisation occurs when the nuclei of the sperm and egg cell fuse.

3. **Fertilisation and development**
   Fertilisation is the fusion of the nucleus of a male sperm cell with the nucleus of the female ovum to form a zygote. During **sexual intercourse** (copulation), the erect **penis** of the male is placed in the **vagina** of the female and with **ejaculation** semen, containing the male sperms are released in the vagina.

   Fertilisation takes place high up in the **Fallopian tube** when a **spermatozoan** enters the **secondary oocyte** (in humans). The secondary oocyte divides into an **ovum** (n) with which the **sperm** (n) fuses to form a **zygote** (2n). The nucleus of the zygote contains a combination of the genes of the two parents, originating from the nucleus of the ovum and sperm. The zygotes now **cleave** (divide) by **mitosis** to form a solid ball of 8 to 16 cells, the **blastomeres**.

   The whole structure, which is moving down the Fallopian tube, is known as a **morula**. Further divisions create a hollow sphere or **blastula**. The blastula develops into a **blastocyst**, which is **implanted** into the **endometrium** of the uterus with its outer layer, the **trophoblast**, and develops into an **embryo**. Where the trophoblast makes contact with the inner wall of the uterus, the **endometrium**, the **placenta** develops.

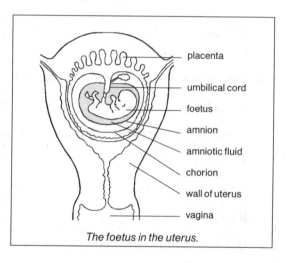

The foetus in the uterus.

4. **Pregnancy**
   The **placenta** is attached to the **embryo** by the **umbilical cord**. Nutrients, oxygen (O_2) and **hormones** are supplied to the developing embryo by the placenta, by diffusion from the blood of the mother. **Carbon dioxide** (CO_2) and **wastes** diffuse from the blood of the embryo into that of the mother. The placenta also produces the hormone, **progesterone**, that helps to keep the endometrium of the uterus in place.

   The embryo is surrounded by **extra-embryonic membranes**. The outer membrane, the **chorion**, has the **amnion** to its inside enclosing the **amniotic fluid**. This cushions the embryo, allows free **movement, symmetrical development** and **protects** it against **disiccation, shock** and **temperature changes**. The **allantois** and **yolk sac** grow from the wall of the alimentary canal of the **foetus**. In humans, these two structures do not function anymore.

   After approximately 280 days, **contractions** of the uterus begin and the fully developed **baby** is pushed through the **birth canal** to the outside. **Labour** is induced by the hormone **oxytocin**, secreted by the hypophysis.

5. **Parental care after pregnancy**
   The baby is fed by the mother on **milk**, secreted by the **mammary glands**. The production of milk is stimulated by the hormone **prolactin**. At the start the milk is known as **colostrum**. It is a laxative and helps the baby to rid itself from bile that accumulated during the foetal stages in the alimentary canal. Milk contains anti-bodies that give the baby passive immunity against certain diseases. Later the baby is fed a balanced diet to prevent malnutrition. The parents are responsible for the baby's care until the child can be self-sufficient.

STANDARD GRADE

QUESTIONS

Section A

A. *Various possibilities are suggested as answers to the following questions. Indicate the correct answer.*

1. The human testes are protected by the
   - A scrotum.
   - B prostate gland.
   - C pericardium.
   - D epididymis.

2. Human sperms are produced in the
   - A seminal vesicles.
   - B prostate gland.
   - C vas deferens.
   - D testes.

3. Before copulation human sperms are temporarily stored in the
   - A sperm ducts.
   - B scrotum.
   - C epididymis.
   - D seminal vesicle.

4. A gland(s) whose secretion helps to activate the sperm is/are the
   - A cells of Sertoli.
   - B vas deferens.
   - C testes.
   - D prostate gland.

5. The process whereby two parent organisms give rise to a new living individual is known as
   - A reproduction.
   - B pollination.
   - C fertilisation.
   - D pregnancy.

6. Which of the following, concerning the number of chromosomes of humans, is wrong?
   - A Zygote = 2n
   - B Embryo = 2n
   - C Gonads = n
   - D Sexual cell = n

7. Which of the following is the odd one out?
   - A Uterus
   - B Ovary
   - C Ovum
   - D Testes

Questions 8 to 11 refer to the following diagram of the female reproductive system.

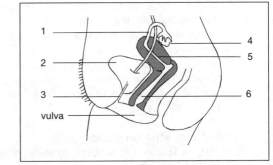

8. Ova are produced in
   - A 1.
   - B 2.
   - C 4.
   - D 5.

9. Fertilisation of the human ovum occurs in
   - A 1.
   - B 2.
   - C 5.
   - D 6.

10. The embryo develops in
    - A 1.
    - B 2.
    - C 5.
    - D 6.

11. Sperms are released in
    - A 1.
    - B 3.
    - C 5.
    - D 6.

12. Which of the following pairs indicate a reproductive structure and its function correctly?
    A   Fallopian tube – production of sperms   C   Uterus – development of embryo
    B   Vagina – fertilisation   D   Testis – production of ovum

13. The fusion of a sperm and ovum is known as
    A   fertilisation.   C   regeneration.
    B   ovulation.   D   cleavage.

14. The stage in the development of the hyman embryo, resembling a mulberry, is called the
    A   foetus.   C   morula.
    B   zygote.   D   blastula.

15. In humans fertilisation normally occurs in the
    A   vagina.   C   uterus.
    B   ovary.   D   Fallopian tube.

16. Fertilisation occurs when the
    A   sperm penetrates the ovum.
    B   sperm makes contact with the ovum.
    C   nucleus of the sperm fuses with the nucleus of the ovum.
    D   fertilisation membrane has formed around the ovum.

17. If the cell of the human possesses 23 chromosomes, it is
    A   diploid.   C   a somatic cell.
    B   a zygote.   D   a gamete.

18. The mammalian embryo develops before birth in the
    A   vagina.   C   uterus.
    B   ovary.   D   ureter.

19. The liquid that surrounds the human embryo is the . . . fluid.
    A   semen   C   tissue
    B   chorion   D   amniotic

20. The developing embryo of the human is surrounded by a sac filled with a fluid which acts as a
    A   shock absorber.
    B   medium which removes excretory products.
    C   medium from which oxygen is obtained.
    D   medium in which secretions take place.

B. Write down the correct term for each of the following.

1. The organ which is used to transfer the sperm from the male to the female
2. A gland of the male reproductive system which secretes a liquid that is added to semen to activate sperms.
3. The liquid secreted by the testes and associated glands containing sperm cells
4. The release of a ripe ovum from the ovary of a female mammal
5. The remainder of the Graafian follicle, after ovulation, in the female ovary
6. The canal leading from the uterus to the exterior, in which the male organ is placed during copulation
7. The hollow organs in mammals in which the embryo develops
8. A tissue complex, richly supplied with maternal and embryonic blood vessels, found in the uterus of a female mammal
9. The small hollow sphere that may become implanted in the wall of the uterus where it develops into a foetus
10. The innermost membrane which protects the embryo in the uterus of a mammal
11. The outer membrane that encloses the embryo in the uterus
12. The intra-uterine period of development of an embryo between fertilisation and birth

C. *Write down the letter of the description in column B which best suits the term in column A*

Column A		Column B
1. Uterus	A	The external opening of the vagina
2. Fallopian tube	B	Secretes milk to nourish the young
3. Testis	C	Is located at the base of the brain
4. Penis	D	Where a mammalian foetus develops
5. Mammary glands	E	An organ enclosed by the scrotum
	F	Deposits spermatozoa in the female body
	G	Transports eggs from the ovary

Section B

1. The diagram alongside represents the reproductive system of a human male.
   (a) Identify the structure numbered 2. What is its function?
   (b) Identify the structure numbered 6. What is its function?
   (c) Identify the gland numbered 5. Which hormone is secreted by this gland?
   (d) Identify the parts numbered 1, 3, 4, 7, 8, 9, 10 and 11.
   (e) Write down the number of the organ in which meiosis occurs. Which cells are produced as a result of this division?

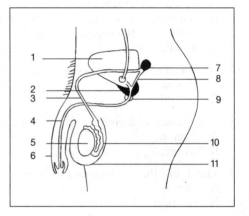

2. The diagram alongside represents the female reproductive system.
   (a) Write down the **number** of the organ in which egg cells are produced. By what type of cell division are egg cells produced?
   (b) Write down the **number** of the organ in which fertilisation usually occurs.
   (c) Write down the **number** and **name** of the organ in which the embryo and foetus develops.
   (d) Name the membranes that enclose the developing foetus. What are the functions of the fluid which surrounds the foetus?
   (e) Which structure joins the foetus to the placenta?
   (f) Name the tube numbered 1.
   (g) Name the process by which an ovum is released into the abdominal cavity.
   (h) What type of reproduction is found in humans? Is it oviparous, ovoviviparous or viviparous? Give a reason for your answer.

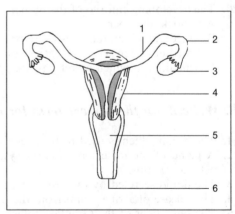

3. The following concerns the reproduction of man.
   (a) Where are: (i) male gametes, and
                  (ii) female gametes produced?
   (b) Where in the body does fertilisation take place?
   (c) Describe briefly the protection and nutrition of the embryo during its development.

4. The diagram alongside represents the uterus of a human female.
   (a) Write down the number of the part which represents the:
      (i) amnion, and
      (ii) chorion.
   (b) What is found in the cavity between the embryo and the membrane, numbered 3?
      State the functions of this substance.
   (c) Identify the part numbered 1.
      Of which tissue is it composed?
      State one function of this tissue.
   (d) Identify the structure numbered 4.
      What is its function?
   (e) Write down the number of the organ which possesses muscles which will push the foetus to the exterior during birth.

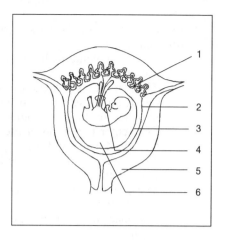

STANDARD GRADE
ANSWERS

Section A

A. 1. A 2. D 3. C 4. D 5. A 6. C 7. D
    8. C 9. A 10. C 11. D 12. C 13. A 14. C
    15. D 16. C 17. D 18. C 19. D 20. A

B. 1. Penis   2. Prostate   3. Semen
    4. Ovulation   5. Corpus luteum   6. Vagina
    7. Uterus   8. Placenta   9. Blastula
    10. Amnion   11. Chorion   12. Pregnancy/Gestation

C. 1. D   2. G   3. E   4. F   5. B

Section B

1. (a) Prostate gland; secretes a liquid which activates sperm; and neutralises acids in the vagina
   (b) Penis; copulating organ; carrying sperms to the vagina of the female
   (c) Testis; testosterone
   (d) 1. Bladder   3. Vas deferens   4. Urethra   7. Seminal vesicle   8. Ejaculatory tube
       9. Cowper's gland   10. Epididymis   11. Scrotum
   (e) no. 5; sperms

2. (a) No. 3; meiosis (b) No. 2
   (c) No. 4; uterus (d) Amnion and chorion; allows free movement and symmetrical development of the embryo; as well as protection against shock and temperature changes
   (e) Umbilical cord   (f) Fallopian tube   (g) Ovulation
   (h) Sexual; viviparous; the young are born alive

3. (*a*) (i) Testes   (ii) Ovaries   (*b*) Fallopian tube
   (*c*) *Nutrition:*
   The ovum contains little yolk and is temporarily fed by nourishing juices; secreted by the uterine wall.
   The placenta is formed for nourishing the embryo; and the umbilical cord allows blood to and from the embryo; so that water, carbohydrates, proteins, fats, vitamins and oxygen; may be carried by the blood of the mother; to the blood spaces of the placenta; and from there enter the blood of the foetus by diffusion
   Carbon dioxide and waste products; are removed in the opposite direction
   The placenta is responsible for nutrition, gaseous exchange and excretion
   Protection:
   The amniotic fluid protects the embryo against shock, dehydration and changes in temperature
   The placenta protects the foetus and acts as a micro-filter for solids and pathogenic organisms
   It allows antibodies to pass to the embryo and provides immunity against infectious diseases

4. (*a*) (i) No. 3   (ii) No. 2
   (*b*) Amniotic fluid; allows free movement and symmetrical development; protects the embryo against shock, dehydration and changes in temperature
   (*c*) The placenta; is the maternal and embryonic nutritive tissue; and is also responsible for nutrition, gaseous exchange and excretion of the foetus
   (*d*) Umbilical cord; transports nutrients; and oxygen to the foetus; and waste products and carbon dioxide; to the blood sinuses of the placenta
   (*e*) No. 5

HIGHER GRADE
QUESTIONS

Section A

A. *Various possibilities are suggested as answers to the following questions. Indicate the correct answer.*

1. Production of the hormone testosterone is a function of the
   A   cells of Leydig.
   B   cells of Sertoli.
   C   Graafian follicles.
   D   epididymis.

   Questions 2 to 4 refer to the schematic representation of spermatogenesis in the human

2. The division that takes place at A is
   A   mitosis.
   B   meiosis.
   C   first meiotic division.
   D   second meiotic division.

3. If there are 46 chromosomes in *a*, how many chromosomes will you expect to find in b_1 or b_2?
   A   92
   B   46
   C   23
   D   46 pairs

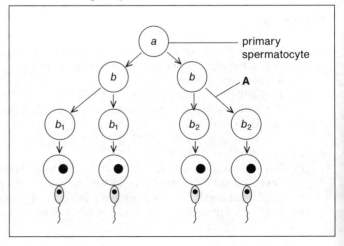

4. In which specific part of the human body does the process take place?
   A   Epididymis
   B   Vas deferens
   C   Cells of Leydig
   D   Lumen of sperm duct

5. Menstruation starts when the production of
   A   oestrogen and progesterone decreases.
   B   progesterone is at its maximum.
   C   oestrogen is at its maximum.
   D   oxytocin decreases.

6. The release of a ripe ovum from the ovary of a female is known as
   A   ovulation.
   B   gametogenesis.
   C   oogenesis.
   D   menstruation.

7. The fusion of a sperm and egg cells is known as
   A   copulation.
   B   cleavage.
   C   fertilisation.
   D   ovulation.

8. In the development of the mammalian embryo, the purpose of the amnion is to
   A   serve as a reserve food supply.
   B   give rise to the placenta.
   C   prevent the developing foetus from moving about.
   D   enclose a fluid which protects the embryo against injury.

9. The hormone responsible for the development of the placenta in mammals is
   A   oestrogen.
   B   adrenalin.
   C   testosterone.
   D   FSH.

10. Progesterone
    A   delays ripening of sperms.
    B   prepares the uterine wall for implantation of the blastocyst.
    C   speeds up the development of follicles.
    D   brings about the formation of the corpus luteum.

11. Contraception pills work effectively because the hormone progesterone in them
    A   stops the development of egg cells.
    B   prevents the thickening of the endometrium wall.
    C   impedes the movement of sperms in the Fallopian tube.
    D   increases the movement of sperms in the Fallopian tube.

12. The main reason why large amounts of male gametes can be formed in the male mammal, is that the
    A   male possesses two testes.
    B   male gametes are formed outside the body of the male.
    C   gametes are formed inside a network of seminiferous tubules.
    D   a high level of testosterone is maintained.

B. Write the correct term for each of the following.

1. The duct or tube leading from the testes to the urethra in males
2. A small organ at the tip of the vulva, sensitive to sexual stimulation
3. A fluid produced by the testes and related glands, containing spermatozoa
4. The part of the male sexual system where sperms undergo maturation
5. The structure in which fertilisation normally takes place in humans
6. The structure at the tip of the sperm cell which makes contact with the egg cell during fertilisation, containing enzymes concerned with fertilisation.
7. A glandular body formed from the remnant of the Graafian follicle after ovulation
8. The release of a secondary oocyte from the ovary into the abdominal cavity
9. The diploid cell in the follicle of the ovary which forms the ovum after meiosis

10. A stage in the development of animals in which the embryo consists of a layer of cells surrounding a cavity
11. A tissue complex, richly supplied with blood vessels from both the mother and the embryo, which is found in the uterus of the female mammal
12. The embryonic membrane which surrounds the other protective membranes of the embryo of a mammal and helps with the formation of the placenta
13. The extra-embryonic membrane which includes the yolk of the egg and is greatly reduced in humans
14. Connecting tissue between the embryo and placenta in humans
15. The monthly cyclic secretion of blood and lining of the uterus when fertilisation is incomplete
16. The removal of a short part of the vas deferens as a contraceptive method

C. Write down the letter of the description in column B which best suits the term in column A.

Column A		Column B
1. Primary follicle	A	Produce testosterone
2. Allele	B	Plays a role in nutrition of spermatozoa
3. Locus	C	Genetic composition of an organism
4. Corpus luteum	D	Position of a gene on a chromosome
5. Cells of Sertoli	E	Changes into a Graafian follicle on maturation
6. Phenotype	F	Secretes progesterone
7. Testosterone	G	Corresponding position of a gene on homologous chromosomes
8. LH	H	External appearance of an organism
	I	Necessary for ovulation
	J	Responsible for male characteristics at puberty
	K	Promotes the development of the luteal body

Section B

1. The diagram alongside shows a human reproductive cell.
   (a) Identify the parts numbered 1 to 5.
   (b) The part numbered 1 contains enzymes. What function is performed by these enzymes during fertilisation?
   (c) The part numbered 3 is packed with mitochondria. Why is it so?
   (d) The ovum (egg) is much larger than the spermatozoon. Why the difference?
   (e) Why is the egg of the pigeon much larger than that of the human?

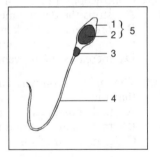

2. The diagram alongside is that of the human female reproductive system. Study it and answer the following questions.
   (a) Identify the parts numbered 1 to 6.
   (b) Fertilisation usually takes place at B. Why does a blockage at A
      (i) prevent fertilisation at B, and
      (ii) not necessarily lead to total infertility?
   (c) Write down the number of the structure lined with
      (i) ciliated epithelium, and
      (ii) endometrium.

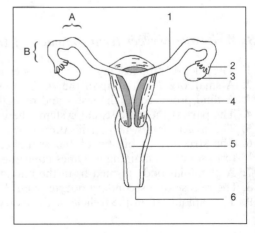

3. (a) Explain how the ovum of a human being develops from the primary oocyte.
   (b) Why will the ovum degenerate if it is not fertilised?
   (c) What is the round mass of cells which develops from the zygote called?
   (d) Name the structure which develops from the mass of cells mentioned in (c).
   (e) Where in the body do these changes mentioned in (c) and (d) take place?

4. The graph below shows the approximate concentration of progestrone and oestrogen in the blood of a mature human female and the thickness of the endometrium over a 28-day period. Use the graph to answer the following questions.

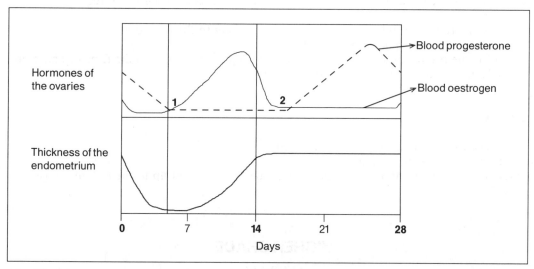

(a) Give the cause of the regular monthly bleeding known as menstruation.
(b) Another hormone "y", causes a sudden rise in blood oestrogen level at point 1 round about day 6. Name "y" and state where it is secreted.
(c) What causes "y" to start work and causing a rise in the level of blood oestrogen?
(d) What changes occur in the ovary at the same time that a drop in oestrogen occurs at 2 on the graph?
(e) Why does the level of blood progesterone begin to rise steeply after point 2 on the graph?
(f) According to this graph on which days, could fertilisation occur?
(g) Describe the role of LH in the female sexual cycle.
(h) Explain why, if the corpus luteum should continue to grow after fertilisation has occurred, no new follicles will be produced.

The diagram represents a stage in the development of a mammalian foetus. Explain the role each of the numbered parts plays in the development of the foetus. In your answer use these **numbers** with their corresponding **names**.

Describe the role played by hormones in the reproductive cycle in a human female.

Answer the following questions on the reproduction of humans.
(a) Name the functions of the following in the testes:
   (i) cells of Sertoli,
   (ii) cells of Leydig (interstitial cells),
   (iii) epididymis, and
   (iv) seminal vesicle.

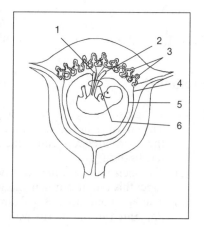

(b) Name the process by which an ovum is released from the ovary.
(c) Where does fertilisation of the ovum take place?
(d) Which structures perform the following functions during the development of the embryo?
   (i) The supply of oxygen
   (ii) The removal of waste products
   (iii) Protection against desiccation
   (iv) Acting as a micro-filter for germs

8. Outline the process whereby a human foetus receives nourishment from the mother and how the waste products are eliminated. (A description of the foetus itself is not required).

9. Discuss briefly the reproduction of the human referring to the provision of
   (a) suitable sites for male gamete development,
   (b) adequate protection for the male gametes during their transfer to the female gamete, and
   (c) adequate nutrition for the developing embryo.

Section C

Write an essay on the reproduction of man with reference to
(a) the formation of gametes, and
(b) internal fertilisation, implantation and embryonic development up to the blastula stage.

HIGHER GRADE
ANSWERS

Section A

A. 1. A   2. D   3. C   4. D   5. A   6. A   7. C
    8. D   9. A   10. B   11. A   12. C

B. 1. Vas deferens   2. Clitoris   3. Semen
    4. Epididymis   5. Fallopian tube   6. Acrosome
    7. Corpus luteum   8. Ovulation   9. Primary oocyte
    10. Blastula   11. Placenta   12. Chorion
    13. Yolk sac   14. Umbilical cord   15. Menstruation
    16. Vasectomy

C. 1. E   2. G   3. D   4. F   5. B   6. H   7. J   8. K

Section B

1. (a) 1. Acrosome   2. Nucleus   3. Neck   4. Tail   5. Head
   (b) The enzymes in the acrosome digest the plasmalemma of the ovum; so that the nucleus of the spermatozoon can penetrate it; to fuse with the nucleus of the ovum and effect fertilisation
   (c) The spermatozoon has to travel (swim) from the vagina through the uterus and Fallopian tube; this requires much energy; which is supplied by ATP; produced in the mitochondria
   (d) The egg contains a small amount of yolk for the early development; until it becomes implanted in the uterus wall

(e) Pigeons are oviparous and the eggs hatch outside the body of the mother in a nest and are covered with a shell
The embryo has to be supplied with food in the form of large amounts of yolk and albumen inside the shell of the egg

2. (a) 1. Fallopian tube 2. Funnel with fimbriae 3. Ovary
4. Uterus 5. Vagina 6. Opening of vagina (vulva)
(b) (i) The sperms cannot reach the ovum
(ii) The egg produced in the other ovary can still be fertilised; every second month
(c) (i) No. 1   (ii) No. 4

3. (a) It takes place in the Graafian follicle of the ovary
The primary oocyte is diploid (2n); and undergoes the first meiotic division; to form a haploid (n) secondary oocyte; and a first polar body
Both these undergo the second meiotic division; during fertilisation only if the ovum is fertilised; the second polar body is formed which degenerates gradually
(b) During the second meiotic division, the centrosome disappears; and no further mitosis can take place
(c) Morula
(d) Blastosist
(e) Fallopian tube and uterus

4. (a) A sudden decrease of the progesterone and oestrogen levels of the blood
(b) FSH: hypophysis
(c) When the endometrium starts to disintegrate
(d) Ovulation
(e) The Graafian follicle; then changes into a corpus luteum; producing progesterone
(f) Between days 12 and 15
(g) LH induces ovulation; and controls the formation of a corpus luteum from the Graafian follicle; producing progesterone
(h) Progesterone produced by the corpus luteum; inhibits the formation of FSH; which controls the development of the Graafian follicle and eventually ovulation

5. 1. – *Allantois*: Forms part of the umbilical cord; plays a role in early blood production fuses with the chorion to form the placenta
2. – *Yolk sac*: Filled with secretions of the uterine glands; no function in humans
3. – *Chorion villi*: Form part of the placenta; diffusion of respiratory gases, food materials and metabolic wastes takes place between the blood of the foetus and that of the mother. It also acts as a micro-filter; preventing pathogenic substances to pass; antibodies are allowed through
4. – *Chorion*: Placental villi develop and extend into blood sinuses of the endometrium; it forms part of the placenta; plays a role in nutrition of the foetus; exchange of materials takes place between the embryonic and maternal blood
5. – *Amnion*: Cavity filled with amniotic fluid; it allows free movement of the embryo and symmetrical development; it also protects the embryo against desiccation and temperature changes
6. – *Umbilical cord*: Two arteries carry blood from the foetus to the placenta; it transports waste products and carbon dioxide. The umbilical vein carries blood from the placenta to the foetus; and transports food materials and oxygen

6. Hypophysis produces **FSH**; it stimulates primary follicle development; the lining of the follicle secretes **oestrogen** which is responsible for secondary sexual characteristics; it also thickens the uterine lining during the ovarian cycle
After ovulation the corpus luteum develops; and its development is controlled by **LH** secreted by the hypophysis

Corpus luteum secretes **progesterone** and it vascularises and glandularises the uterine wall; and inhibits production of **FSH** and prevents further follicle formation
If no fertilisation takes place the corpus luteum degenerates; and the **progesterone** production decreases or ceases
FSH is now secreted and stimulates another egg to ripen; and the endometrium degenerates and sloughs off (menstruation)

7. (a) (i) Nutrition of spermatozoa
(ii) Produce testosterone
(iii) Storage and maturation of spermatozoa
(iv) Secrete fructose as source of food for spermatozoa; and other secretions which keep spermatozoa motile
(b) Ovulation
(c) Fallopian tube
(d) (i) Placenta (ii) Placenta (iii) Amnion and amniotic fluid (iv) Placenta

8. Food is absorbed by the maternal gut; and diffuses into the maternal bloodstream
The maternal and foetal circulations are separate; and the placenta interdigitates with the uterine wall; and foodstuffs diffuse from the maternal to the foetal bloodstream; along the umbilical cord in a foetal vein to the foetus
Waste products move along the foetal artery into the umbilical cord; and from the placenta they diffuse into the maternal bloodstream. Urea is excreted by the maternal kidneys; and carbon dioxide is excreted by maternal lungs

9. (a) A pair of testes; outside the trunk; in scrotum; creates a low temperature for spermatogenesis. Many spermatogonia; in seminiferous tubules; produce sperms
(b) The penis; transports sperm to the vagina
Sperm is in a seminal fluid; and swims up in the Fallopian tube
The semen is alkaline; as a result of the secretion of the prostate gland; and it neutralises acids in the vagina
A small amount of yolk; and nutritional juices of the uterus feed the embryo before implantation
(c) The chorion, develops villi in blood spaces; of the uterine wall; for the exchange of nutrients; between the blood of the embryo and blood of the mother, the placenta

Section C

(a) The male gametes are produced in the testes; by a process known as **spermatogenesis**. The germinal epithelium (gonocytes) of the seminiferous tubules form new cells, the spermatogonia (2n) by mitosis; and they each mature into a primary spermatocyte (2n); which now undergo meiosis (reduction division); to form spermatids (n). Each spermatid matures into a spermatozoon which is stored in the epididymis; until required. They are supplied with nutrients by the cells of Sertoli
The female gametes are produced in the ovaries by a process known as **oogenesis**. The primary follicle in the ovary contains a single oogonium (2n); and during the productive phase it changes into a primary oocyte; as a result of the influence of FSH; of the hypophysis. Such a ripe follicle is a Graafian follicle; and it moves to the surface of the ovary; ripens; swells up and splits open and is set free (ovulation).
Meiosis now begins, the **first division** produces two haploid cells. The one with the most cytoplasm is called the secondary oocyte; and the one without cytoplasm, the polar body. The polar body disappears; and further development only occurs after fertilisation
(b) Spermatozoa are deposited by the penis; during sexual intercourse, into the vagina of the female. The spermatozoa move (swim) through the uterus; up the Fallopian tube and reach the secondary oocyte (egg cell) released during ovulation.

The acrosome of the spermatozoon secretes certain enzymes which digest the outer membrane of the oocyte. The nucleus of the spermatozoon enters the ovum and the ovum forms a fertilisation membrane; which prevents the entrance of any other spermatozoa.

The secondary oocyte (egg) undergoes the **second meiotic division**; and forms and ovum and second polar body (which disappears). The nucleus (n) of the spermatozoon; fuses with the nucleus (n) of the ovum; to form a diploid zygote (2n).

By peristaltic contractions, the fertilised cell moves down the Fallopian tube; and splits into blastomeres; which is a small round ball of cells; called the **morula**. The morula keeps on dividing into a **blastula** and the outer layer of the **blastocyst** (as it is now known) forms the **trophoblast** out of which the placenta develops when it implants into the endometrium of the uterus.

Notes: